W0049821

# symposia on theoretical physics and mathematics

**7**

*Contributors to this volume:*

P. L. Kannappan
A. N. Mitra
F. Pham
G. Rickayzen
M. Scadron
S. K. Singh
S. K. Srinivasan
B. M. Udgaonkar

# symposia on theoretical physics and mathematics

Lectures presented at the
1966 Summer School
of the Institute
of Mathematical Sciences
Madras, India

Edited by
**ALLADI RAMAKRISHNAN**
Director of the Institute

PLENUM PRESS • NEW YORK • 1968

ISBN-13: 978-1-4684-7729-0     e-ISBN-13: 978-1-4684-7727-6
DOI: 10.1007/ 978-1-4684-7727-6
Library of Congress Catalog Card Number 65-21184

©1968 Plenum Press
Softcover reprint of the hardcover 1st edition 1968
A Division of Plenum Publishing Corporation
227 West 17 Street, New York, N. Y. 10011

All rights reserved

No part of this publication may be reproduced in any
form without written permission from the publisher.

# Introduction

This volume contains the proceedings of the Third Matscience Summer School held at Bangalore in September, 1966. The special feature of these proceedings was two systematic series of lectures, one by F. Pham of C.E.N., Saclay and CERN, Geneva and the other by G. Rickayzen of the University of Kent, Canterbury.

Pham dwelt at length on the applications of the methods of algebraic topology and differential forms to the study of the analytic properties of $S$-matrix theory, in particular, with reference to the location of singularities of the multiple scattering processes. This exposition was a natural sequel to the lectures of V. L. Teplitz, published in an earlier volume of this series.

Rickayzen discussed in detail the latest theory of superconductivity. Other lectures were those of Scadron, who dealt with some formal features of potential scattering theory, and B. M. Udgaonkar and A. N. Mitra, who spoke on certain aspects of bootstraps and quark models, respectively.

The contributions in pure mathematics in this volume include two lectures by S. K. Singh, one on the field of Mikusinski operators and another on Riemann mapping theorem, and a lecture on cosine functionals by P. L. Kannappan.

One of the highlights of the symposium was a lecture by S. K. Srinivasan who is keeping alive the interest of the Madras group in the theory of stochastic processes and who, in particular, has enlarged the domain of the application of the theory of product densities.

*Alladi Ramakrishnan*

# Contents

# Contents of Other Volumes

# VOLUME 3

# VOLUME 4

# VOLUME 5

# VOLUME 6

# VOLUME 8

# Superconductivity

G. RICKAYZEN

*UNIVERSITY OF KENT AT CANTERBURY*
*Canterbury, England*

---

## 1. INTRODUCTION

I wish to present a review of our present understanding of the phenomenon of superconductivity. It was recognized in the early work of London and others that superconductivity is a cooperative effect of the electrons such that they condense into a many-body state described by a single wave function and this state is such that it resists deformations. The concept of such a cooperative effect is inherent in the wave function of Bardeen, Cooper, and Schrieffer[1] (referred to hereafter as BCS) where the Bloch states are paired, the different pairs being both occupied or unoccupied, in the configurations which are coherently superposed in the many-body wave function.

Landau[2] attempted to account for the rigidity of the wave function in terms of the low-lying excitations of the system. He was concerned with superfluidity but his analysis is easily carried over to superconductivity. If we suppose the system is uniform and isotropic, then in the ground state it will possess excitations with momentum $\mathbf{p}$ and energy $\epsilon(\mathbf{p})$. If the whole system is given a velocity $\mathbf{v}$, there will be a scattering mechanism which will tend to restore the system to equilibrium in the original rest frame. By Galilean invariance, in this frame the energy of the excitations is $\epsilon(\mathbf{p}) - \mathbf{p} \cdot \mathbf{v}$. If this is positive, electrons will remain in the condensate and

1

superconductivity will be maintained. If it is negative, electrons will leave the condensate to become excitations and superconductivity will tend to be destroyed. Hence, Landau's criterion for superconductivity is that $\epsilon(\mathbf{p}) - \mathbf{p} \cdot \mathbf{v}$ should be positive for all $\mathbf{p}$. In the worst case of $\mathbf{p}$ parallel to $\mathbf{v}$, this requires

$$\frac{\epsilon(\mathbf{p})}{|\mathbf{p}|} > \mathbf{v}$$

In a superconductor $\mathbf{p} \approx \mathbf{p}_F$, the Fermi momentum. Hence, the criterion for superconductivity is that there should be no excitations of infinitely small energy, that is, there should be a gap in the spectrum of excitations. In the theory of BCS, there is indeed a gap and Landau's criterion is satisfied.†

It is now generally realized, however, that although Landau's criterion may be sufficient for superconductivity it is by no means necessary. Both theory and experiment now confirm that it is possible to have gapless superconductivity in, for example, thin films in a high parallel field[3] in a type-II superconductor when the field lies between $H_{c_1}$ and $H_{c_2}$[4] and in a superconductor containing a sufficiently high concentration of paramagnetic impurities.[5] Even in a BCS-type superconductor at a finite temperature, Landau's criterion is broken because there are always phonons present to excite electrons from the condensate. In fact, this is strictly another case of gapless superconductivity.[6] Hence, in all the cases of practical importance the criterion is not valid. How is this possible? It is possible because, although the criterion shows when electrons will start to leave the condensate to form thermal excitations it does not show that all electrons will leave the condensate. In fact, electrons continue to leave the condensate until positive energy (or free energy) is required to create more. As long as some electrons remain in the condensate, superconductivity continues. Hence, Landau's criterion gives only the velocity at which some electrons leave the condensate, that is, it gives the velocity at which the density of superconducting electrons $n_S$, becomes less than the density of conduction electrons $n$. Since the criterion is broken in all the cases mentioned above, we expect $n_S = n$ only in a pure, homogeneous

---

†Strictly, one cannot apply the argument of Galilean invariance to a superconductor, because of the existence of the lattice. Most theoretical models of superconductivity including that of BCS, however, are Galilean invariant and lead to energy spectra of the form assumed above.

superconductor at the absolute zero. This conclusion is supported by many calculations. (Experiments usually do not measure $n_s$ directly.)

If Landau's condition is not necessary for superconductivity, then what condition is? The answer is given directly in terms of the wave function or density matrix. In simplest terms it is that in a superconductor we have a new macroscopic variable $F(\mathbf{r})$ defined by

$$F(\mathbf{r}) = \langle \psi_{\downarrow}(\mathbf{r})\psi_{\uparrow}(\mathbf{r}) \rangle$$

where $\psi_{\sigma}(\mathbf{r})$ is a field operator for electrons.[†] Here, the average is a quantum and thermodynamic average. If one also averages over small but macroscopic regions of space, one has that for a macroscopically uniform system in the absence of magnetic fields $F(\mathbf{r})$ has the form

$$F(\mathbf{r}) = F(\mathbf{p})e^{i\mathbf{p}\cdot\mathbf{r}}$$

The vector $\mathbf{p}$ represents the flow of the system, and the function $F(\mathbf{p})$ is determined by the condition that the system be in thermodynamic equilibrium subject to the condition that the phase of $F(\mathbf{r})$ be $\mathbf{p}\cdot\mathbf{r}$. In the BCS theory, a function with phase $(\mathbf{p}\cdot\mathbf{r})$ arises from the pairing of $(\mathbf{k} + \mathbf{p}/2, \uparrow)$ with $(-\mathbf{k} + \mathbf{p}/2, \downarrow)$.

The states corresponding to different values of $\mathbf{p}$ for which $F(\mathbf{p})$ is non-zero are such that the excitation energy of an electron in any one of them is positive and the matrix element of any single particle operator between any two of them is zero. Hence, one expects that scattering cannot reduce any one of these states to another; for each $\mathbf{p}$ we have a different state of metastable equilibrium. The condition, then, for superconductivity is that, for small $\mathbf{p}$, $F(\mathbf{p})$ should not be zero.

If this is true, one should be able to prove it. In fact, when impurity scattering is important, this has been shown by direct calculation for a number of different models.[5,8] Impurity scattering, however, is elastic and is, therefore, particularly inefficient for destroying superconductivity. One should show also that inelastic scattering such as scattering by phonons, does not destroy superconductivity. This has been done by considering the connection between

---

[†]This method assumes that the eigenstates do not have a definite number of particles present. An alternative method which allows eigenstates with definite numbers of particles has been given by Yang.[7]

the Meissner effect and the infinite conductivity.[9]† [More recently, this question has also been discussed by L. Pičman (see Ref. 14).] We propose to review this work here. Only the case of a simply connected superconductor will be considered since the problem of the persistent current in a multiply connected superconductor has been considered by Wentzel.[10]

## 2. MACROSCOPIC ANALYSIS

First, we consider what quantities we will have to calculate from the microscopic theory and what conditions they must satisfy in order that the Meissner effect and the infinite conductivity follow. Both these effects are weak field effects so we shall consider only the response of the system which is linear in the applied fields. Further, since both effects occur in transverse fields, we shall consider only the effects of such fields. Since the magnetic field and electric field are not independent in a general transverse field, we need consider only the effect of an electric field.

If the superconductor is also macroscopically uniform, different Fourier components will give rise to independent linear effects. The response of the system will be described by the induced current $j_i(\mathbf{q}, \omega)$ and one component $E(\mathbf{q}, \omega)$ will give rise to the corresponding current $j_i(\mathbf{q}, \omega)$. Now, the only polar vector we can form from $\mathbf{q}$ and a transverse field $\mathbf{E}$ which is linear in $\mathbf{E}$ is $\mathbf{E}$ itself. Hence, the most general possible relation between $\mathbf{j}$ and $\mathbf{E}$ is

$$\frac{4\pi}{c^2} j_i(\mathbf{q}, \omega) = \frac{iK(q, \omega)}{\omega} E(\mathbf{q}, \omega) \tag{2}$$

where $K(q, \omega)$ is an arbitrary function of $\omega$ and of $q$, the modulus of the vector $\mathbf{q}$. The constants which appear in equation (2) are purely conventional.

The response of the system to a uniform field is given by equa-

---

†In a footnote to a recent preprint, P. C. Martin has argued that our considerations are not necessary to show that the Meissner effect and infinite conductivity go together. Once it is known that the system possesses a measureable conductivity in the experimentalist's sense, he argues, the two effects must imply each other. Martin's argument is very general but not obviously foolproof. If it should survive the test of time, then this paper confirms it and shows that according to the BCS theory a superconductor does possess a measureable conductivity.

tion (2) with $\mathbf{q}$ equal to zero. Hence, the frequency dependent conductivity is

$$\sigma(\omega) = \frac{ic^2}{4\pi\omega} K(0, \omega)$$

For infinite conductivity we require that this should tend to infinity as $\omega$ tends to zero. If we are more specific and ask for the electrons to behave at low frequencies like a gas of $n_{S1}$ freely accelerating electrons then we require that as $\omega$ tends to zero,

$$\frac{\partial \mathbf{j}}{\partial t} = \frac{n_{S1}e^2}{m} \mathbf{E}$$

This means that as $\omega$ tends to zero,

$$K(0, \omega) \rightarrow \frac{4\pi n_{S1}e^2}{mc^2} \qquad \text{a positive constant} \qquad (3)$$

To obtain the conditions for the Meissner effect we have to consider the effect of a static magnetic field on the superconductor. If the magnetic induction in the superconductor is $\mathbf{B}$ we have, from Maxwell's equations, that for a general field

$$i\mathbf{q} \times \mathbf{E} = \frac{i\omega\mathbf{B}}{c}$$

If the field is transverse, this can be rewritten

$$\mathbf{E} = -\frac{\omega\mathbf{q} \times \mathbf{B}}{cq^2}$$

and the relation between induced current and magnetic field is

$$\frac{4\pi}{c} \mathbf{j}_t(q, \omega) = -\frac{iK(q, \omega)}{q^2} (\mathbf{q} \times \mathbf{B})$$

In the limit of static fields, this is

$$\frac{4\pi}{c} \mathbf{j}_t(\mathbf{q}, 0) = -\frac{iK(\mathbf{q}, 0)}{q^2} (\mathbf{q} \times \mathbf{B})$$

If a source $\mathbf{j}_s$ is present, we have that $\mathbf{B}$ is determined by

$$\text{curl } \mathbf{B} = \frac{4\pi}{c} (\mathbf{j}_t + \mathbf{j}_s)$$

For a given source $\mathbf{j}_s$, these equations can be solved for the induction $\mathbf{B}$. For the special case of a long, thin, superconductor parallel to a uniform external field $H$ with specular reflection of the electrons at the boundary, the solution is easily obtained.[11] The induction at a distance $z$ from the surface is

$$B(z) = -\frac{iH}{\pi} \int_{-\infty}^{+\infty} dq \, \frac{q e^{iqz}}{q^2 + K(q, 0)}$$

The behavior of $B$ at large distances from the surface is determined by the behavior of the integrand for small $q$. If we require that the field decrease exponentially with distance at large distances from the surface, then we must have

$$\lim_{q \to 0} K(q, 0) = \text{positive constant}$$

By analogy with equation (3), we shall write the constant as $4\pi n_s e^2 / mc^2$. Hence, for a Meissner effect in the above sense,

$$\lim_{q \to 0} K(q, 0) = \frac{4\pi n_{S_2} e^2}{mc^2} \tag{4}$$

We see then that the two effects depend on the behavior of $K(q, \omega)$ as $q$ and $\omega$ tend to zero, the only difference being in the order in which the limits are taken. One might surmise that the order of the limits is irrelevant and that the two quantities (3) and (4) are always equal. This would correspond with the idea of Section (1) that the current carrying states are in metastable equilibrium. Nevertheless, it is possible to construct a trivial example where the limits are not equal. This is the case of a free electron gas for which all the electrons are freely accelerate and so $n_{S_1}$ is equal to $n$, whereas there is no Meissner effect and $n_{S_2}$ is zero. However, this case includes no scattering to bring the system into equilibrium with its surroundings. We might guess, therefore, that when such scattering is present, the two limits are equal and the phenomena are equivalent. This is the approach of Ref. 9 which we review here.

## 3. FORMALITIES

We have shown that the effects depend on the one function $K(q, \omega)$ which, to determine whether or not the effects exists, must be calculated from the appropriate microscopic theory. We shall delay introducing a specific theory for as long as possible, deriving instead formal expressions for $K(q, \omega)$ in terms of microscopic quantities. In this way we shall be able to pinpoint the condition necessary for the two phenomena to occur together.

For the Meissner effect we require $K(q, 0)$, which links the current density to a static magnetic field when the system is in thermal equilibrium. If $\mathcal{H}$ is the Hamiltonian in the absence of the field, the Hamiltonian in the presence of the field, to first order in the vector potential, is

$$\mathcal{H} + \mathcal{H}_A$$

where

$$\mathcal{H}_A = -\frac{1}{c} \int d^3r \, \mathbf{A} \cdot \mathbf{j}_{op} \tag{5}$$

and

$$\mathbf{j}_{op} = \frac{eh}{2im} [\psi^+(\mathbf{r})\nabla\psi(\mathbf{r}) - (\nabla\psi^+)\psi] - \frac{e^2}{mc}\psi^+(\mathbf{r})\psi(\mathbf{r})\mathbf{A}(\mathbf{r}) \tag{6}$$

The microscopic current density is then

$$\mathbf{J}(\mathbf{r}) = \frac{\mathrm{tr} \{\exp[-\beta(\mathcal{H} + \mathcal{H}_A)]j_{op}\}}{\mathrm{tr}\exp[-\beta(\mathcal{H} + \mathcal{H}_A)]}$$

By standard techniques of perturbation theory this can be expanded in powers of $\mathcal{H}_A$. To the first order in the vector potential one then has

$$\mathbf{J}(\mathbf{r}) = \mathrm{tr}\,\rho\mathbf{j}_{op} - \int_0^\beta d\tau \,\mathrm{tr}\,[\rho\mathcal{H}_A(\tau)\mathbf{j}_{op}(\mathbf{r})] \tag{7}$$

where the operator $\rho$ is the density matrix for the imperturbed system, that is,

$$\rho = \frac{\exp[-\beta\mathcal{H}]}{\mathrm{tr}\exp(-\beta\mathcal{H})} = \frac{\exp(-\beta\mathcal{H})}{Z} \tag{8}$$

and $\mathcal{H}_A(\tau)$ is a Heisenberg operator defined with an "imaginary" time by

$$\mathcal{H}_A(\tau) = \exp(\mathcal{H}\tau)\,\mathcal{H}_A\exp(-\mathcal{H}\tau) \tag{9}$$

Equation (7) yields a formal expression for the microscopic current density induced by the field. The quantity considered in the previous section is the macroscopic current density. This is to be obtained from the expression (7) by averaging that expression over volumes whose dimensions are large compared with atomic dimensions but small compared with the appropriate wavelength of the field. Since we are interested in the response of the system to very long wavelength fields, this average is always possible. Without indicating this average by a special symbol, we shall assume that it is performed, and then equation (7) yields the macroscopic current density.

Because no current flows when the system is in thermal equilibrium, only the term of $\mathbf{j}_{op}$ which explicitly contains the vector potential contributes to the first term of $\mathbf{J}(\mathbf{r})$. Further, just this part of $\mathbf{j}_{op}$ does not contribute to the second term of $\mathbf{J}(\mathbf{r})$ at least to the first order in $\mathbf{A}$. Hence, $\mathbf{J}(\mathbf{r})$ can be rewritten

$$J_\mu(\mathbf{r}) = -\frac{e^2}{mc}\,\mathrm{tr}\,[\rho\psi^*\psi]A_\mu(\mathbf{r}) - \frac{1}{c}\int_0^\beta d\tau \int d^3\mathbf{r}'\,\mathrm{Tr}\,\rho j_\nu(\mathbf{r}',\tau)j_\mu(\mathbf{r},0)A_\nu(\mathbf{r}',\tau)$$

$$(10)$$

where $j_\mu$ stands for

$$\frac{eh}{2im}\,[\psi^+\nabla\psi - (\nabla\psi^+)\psi]$$

The q-th Fourier component of $J_\mu(\mathbf{r})$ is easily abstracted from this

$$J_\nu(\mathbf{q}) = -\frac{ne^2}{mc}A_\nu(\mathbf{q}) + \frac{1}{c}\int_0^\beta d\tau\,\mathrm{tr}\,\rho j_\nu(-q,\tau)j_\mu(q,0)A_\nu(\mathbf{q},\tau)$$

In deriving this we have also used the fact that $\mathrm{tr}\,(\rho\psi^+\psi)$ is the total density of electrons $n$. The q-th Fourier component of the microscopic current density would be related to other Fourier components of $\mathbf{A}$. For example, if the lattice is included in $\mathscr{H}$, then microscopic $J_\mu(q)$ would be related to $A_\mu(\mathbf{q}+\mathbf{K})$, where $\mathbf{K}$ is a vector of the reciprocal lattice. Such terms oscillate rapidly over atomic dimensions and are removed by the averaging procedure. Similar remarks apply to the effects of impurities. Thus, it is at this stage that we have gone over to the macroscopic theory.

We can perform the $\tau$ integration if we introduce the exact set of eigenstates $|m\rangle$ of $\mathscr{H}$ with their corresponding eigenvalues $E_m$. If the system is isotropic, we can at the same time abstract the response function $K(\mathbf{q},0)$ to find

$$\frac{cK(\mathbf{q},0)}{4\pi} = \frac{ne^2}{mc} + \frac{1}{3c}\sum_{m,n}\frac{e^{-\beta E_m} - e^{-\beta E_n}}{Z}\frac{j_{\mu mn}(\mathbf{q})j_{\mu nm}(\mathbf{q})}{E_m - E_n} \qquad (11)$$

where

$$j_{\mu mn}(\mathbf{q}) = \langle m|j_{\mu op}(\mathbf{q})|n\rangle$$

We will leave this aside for the moment and look for a similar expression for $K(\mathbf{q},\omega)$. This is found from Kubo's formula[12] for the response of the system to a time dependent perturbation, and in our case leads to

$$J_\mu(\mathbf{r}, t) = \text{tr } \rho j_\mu + \frac{i}{\hbar} \int_0^\infty dt' \text{ tr } [\rho[\tilde{\mathscr{H}}_A(t'), j_\mu(\mathbf{r}, t)]]\theta(t - t') \tag{12}$$

where

$$\tilde{\mathscr{H}}_A(t) = \exp\left(\frac{i\mathscr{H}t}{\hbar}\right)\mathscr{H}_A(t)\exp\left(\frac{-i\mathscr{H}t}{\hbar}\right) \tag{13}$$

and

$$\tilde{\mathscr{H}}_A(t) = -\frac{1}{c}\int \mathbf{A}(\mathbf{r}', t)\mathbf{j}_{op}(\mathbf{r}')d^3r'$$

Here $\mathbf{J}(\mathbf{r}, t)$ is the current density at time $t$ when the field $\mathbf{A}(\mathbf{r}, t)$ is switched on at the zero of time.

If $\mathbf{A}(\mathbf{r}, t)$ varies sinusoidally with time as $e^{-i\omega t}$, the part of the current density which varies with time in the same way is

$$J_\mu(\mathbf{r}, \omega) = -\frac{ne^2}{mc}A_\mu(\mathbf{r}, \omega) - \int d^3r' F_{\mu\nu}(\mathbf{r} - \mathbf{r}', \omega)A_\nu(\mathbf{r}', \omega)$$

where

$$F_{\mu\nu}(\mathbf{r} - \mathbf{r}', \omega) = \frac{1}{2\pi}\int_{-\infty}^{+\infty}\frac{1}{c}\text{ tr }\rho j_\nu(\mathbf{r}', t')j_\mu(\mathbf{r}, t)e^{i\omega(t-t')}dt'\,\theta(t - t') \tag{14}$$

Hence, the q-th Fourier component of this current density is given by

$$\mathbf{J}(\mathbf{q}, \omega) = -\frac{ne^2}{mc}\mathbf{A}(\mathbf{q}, \omega)$$

$$- \frac{1}{3c}\int\frac{dt'}{2\pi}\text{ tr }[\rho\tilde{j}_\mu(-\mathbf{q}, t')\tilde{j}_\mu(\mathbf{q}, t)]\theta(t - t')\mathbf{A}(\mathbf{q}, \omega)e^{i\omega(t-t')}$$

If we again introduce the eigenstates of $\mathscr{H}$, we have

$$\frac{cK(\mathbf{q}, \omega)}{4\pi} = \frac{ne^2}{mc} + \frac{1}{3c}\sum_{m,n}\frac{e^{-\beta E_m} - e^{-\beta E_n}}{Z}\frac{j_{\mu mn}(-\mathbf{q})j_{\nu nm}(\mathbf{q})}{E_m - E_n + \hbar\omega + i\delta} \tag{15}$$

The quantity we require to decide whether the conductivity is infinite is

$$\frac{cK(0, \omega)}{4\pi} \approx \frac{ne^2}{mc} - \frac{1}{3c}\sum_{m,n}\frac{e^{-\beta E_m} - e^{-\beta E_n}}{Z}\frac{j_{\mu mn}(0)j_{\nu nm}(0)}{E_m - E_n + \hbar\omega} \tag{16}$$

where $\omega$ is a small complex frequency with positive imaginary part.

The question we have to decide is the following: When is the expression (16) with $\omega$ tending to zero the same as the expression (11) with $\mathbf{q}$ tending to zero? In the case of free particles $j_\mu(0)$ is proportional to the total momentum operator of the system and the states can be chosen so that $j_{\mu m n}(0)$ is diagonal. Then each term of the sum in (16) is zero for $\omega$ not equal to zero and we have

$$\lim_{\omega \to 0} \frac{cK(0, \omega)}{4\pi} = \frac{ne^2}{mc}$$

On the other hand, for finite $\mathbf{q}$, $j_{\mu m n}(\mathbf{q})$ is not quite diagonal. As $\mathbf{q}$ tends to zero $E_m$ tends to $E_n$ in expression (11) and we have

$$\lim_{q \to 0} \frac{cK(q, 0)}{4\pi} = \frac{ne^2}{mc} + \frac{1}{3c} \sum_m \frac{1}{Z} \frac{\partial}{\partial E_m} e^{-\beta E_m} j_{\mu m m}^2(0)$$

The second term is evidently negative and non-zero. Hence, this limit is less than $ne^2/mc$ and can, in fact, be shown to be zero, as it should be.

The function $F_{\mu\nu}(\mathbf{r}, \omega)$ can be defined through equation (14) for complex $\omega$ in the upper half-plane. It is evidently an analytic function of $\omega$ in this region. If it is also analytic in a region about the origin which is independent of $\mathbf{r}$ and $\mathbf{r}'$, then it is clear that the two limits of $K(\mathbf{q}, \omega)$ are equal and the result is proved. This would be the case if after any response is removed, the system returns exponentially fast to equilibrium with a maximum time constant independent of the deformation (when the wavelength of the deformation is large). For then $\mathrm{tr}\,[\rho j_\nu(\mathbf{r}', t')j_\mu(\mathbf{r}, t)]$ would behave like $\exp\left[-(t - t')/\tau\right]$ or less for large $t$ where $\tau$ is independent of $\mathbf{r}$ and $\mathbf{r}'$. However, in general this has not been shown to be the case. For this reason the following argument is not trivial.

## 4. MICROSCOPIC CONDITION FOR THE EQUIVALENCE OF THE MEISSNER EFFECT AND SUPERCONDUCTIVITY

The object of this section is to write the response function in terms of the spectral distribution functions of the vertex part and, hence, to deduce a necessary and sufficient condition for the limiting procedure $\omega \to 0$, $q \to 0$ to be independent of the order in which we take the limits.

We define the vertex function by

$$\Gamma_\mu(\mathbf{k}, \mathbf{k}', \omega_n, \omega_{n'}) = \frac{1}{\beta^2} \int_0^\beta \int_0^\beta d\tau_1\, d\tau_2 \Gamma_\mu(\mathbf{k}, \mathbf{k}', \tau_1, \tau_2) e^{i\omega_n\tau_1 - i\omega_{n'}\tau_2} \qquad (17)$$

and $\Gamma_\mu$ is a matrix in spin space with components

$$\Gamma_{\mu\alpha\beta}(\mathbf{k}, \mathbf{k}', \tau_1, \tau_2) = -\mathrm{tr}\,\rho T \langle c_{k\beta}^*(\tau_1) c_{k'\alpha}(\tau_2) j_\mu(\mathbf{k} - \mathbf{k}', 0) \rangle \qquad (18)$$

where

$$j_\mu(\mathbf{k} - \mathbf{k}', 0) = \frac{1}{2m} \sum_{K\alpha} (2K + k + k') c^*_{K+k',\alpha}(0) c_{K+k,\alpha}(0) \qquad (19)$$

The arguments $\tau$ which appear here are imaginary times and lie between zero and $\beta$.

The vertex function is related to the response function by virtue of the equation

$$\frac{2\beta e\hbar^2}{mv} \sum_k \sum_{\omega_n} Sp k_\nu \Gamma_\mu(\mathbf{k}, \mathbf{k} + \mathbf{q}, \omega_n, \omega_n + \omega_p) = F_{\mu\nu}(\mathbf{q}, i\omega_p) \qquad (20)$$

where $v$ is the volume and

$$F_{\mu\nu}(\mathbf{q}, i\omega_p) = \frac{1}{c} \sum_{m,n} \frac{e^{-\beta E_m} - e^{-\beta E_n}}{Z} \frac{j_{\nu mn}(-\mathbf{q}) j_{\mu nm}(\mathbf{q})}{E_m - E_n + i\omega_p} = \delta_{\mu\nu} F(\mathbf{q}, i\omega_p) \qquad (21)$$

By standard techniques $F(\mathbf{q}, i\omega_p)$ can be analytically continued to the required function $F(\mathbf{q}, \omega)$.

The spectral form of the function $\Gamma(\mathbf{k}, \mathbf{k}', \omega_n, \omega_{n'})$ can be found by inserting complete sets of exact eigenstates of the Hamiltonian between the operators which appear in equation (18) and by then performing the integrations of $\tau_1$ and $\tau_2$ in equation (17). The result is that

$$\Gamma_\mu(\mathbf{k}, \mathbf{k}', \omega_n, \omega_{n'}) = \frac{1}{\beta^2} \iint dy\, dy' \left\{ \frac{J_\mu(\mathbf{k}, \mathbf{k}', y, y')}{(y + i\omega_n)(y' + i\omega_{n'})} \right.$$
$$\left. + \frac{K_\mu(\mathbf{k}, \mathbf{k}', y, y')}{[y + i(\omega_n - \omega_{n'})](y' + i\omega_{n'})} \right\} \qquad (22)$$

where the spectral functions $J$ and $K$ are given by

$$J_{\mu\alpha\beta}(\mathbf{k}, \mathbf{k}', y, y') = \sum_{mnl} \frac{e^{-\beta E_m}}{Z} [(c^*_{k\beta})_{mn}(c_{k'\alpha})_{nl}(j_{\mathbf{k}-\mathbf{k}'\mu})_{lm}(e^{\beta y} + 1)$$
$$\delta(y + E_n - E_m)\delta(y' + E_n - E_l)$$
$$- (c_{k'\alpha})_{mn}(c^*_{k\beta})_{nl}(j_{\mathbf{k}-\mathbf{k}',\mu})_{l,m} e^{-\beta y'}(e^{\beta y} + 1)$$
$$\delta(y + E_l - E_n)\delta(y' + E_m - E_n)] \qquad (23)$$

and

$$K_{\mu\alpha\beta}(\mathbf{k}, \mathbf{k}', y, y') = \sum_{m,n,l} \frac{e^{-\beta E_m}}{Z} [(c^*_{k\beta})_{mn}(c_{k'\alpha})_{nl}(j_{\mathbf{k}-\mathbf{k}',\mu})_{lm}(e^{\beta y} - 1)$$
$$\delta(y + E_l - E_m)\delta(y' + E_n - E_l)$$
$$+ (c_{k'\alpha})_{m,n}(c^*_{k\beta})_{nl}(j_{\mathbf{k}-\mathbf{k}',\mu})_{lm}(e^{\beta y} - 1)$$
$$\delta(y + E_l - E_m)\delta(y' + E_m - E_n) \qquad (24)$$

The integrals of these spectral functions over $y$ and $y'$ are finite and given by

$$\iint dy\, dy'\, J_{\mu\alpha\beta}(\mathbf{k}, \mathbf{k}', y, y') = \operatorname{tr} \rho\{c_{k\beta}^*[c_{k'\alpha}, j_{\mu\mathbf{k}-\mathbf{k}'}] + [c_{k'\alpha}, j_{\mu\mathbf{k}-\mathbf{k}'}]c_{k\beta}^*\}$$

$$= \left(\frac{\mathbf{k}_\mu + \mathbf{k}_\mu'}{m}\right)\delta_{\alpha\beta}$$

$$\iint dy\, dy'\, K_{\mu\alpha\beta}(\mathbf{k}, \mathbf{k}', y, y') = 0$$

We shall be concerned shortly with the singularities of $J_\mu$ and $K_\mu$. The fact that these integrals are finite, limits the possible kinds of singularity.

One can now perform the sum over $\omega_n$ in equation (20) to find $F_{\mu\nu}(\mathbf{q}, i\omega_p)$ in terms of the spectral functions. The result is

$$F_{\mu\nu}(\mathbf{q}, i\omega_p) = \frac{2eh^2}{m}\int dy\, dy' \left\{ J_{\mu\nu}(\mathbf{q}, y, y')\frac{f(y) - f(y')}{y - y' - i\omega_p} \right.$$

$$\left. + \frac{K_{\mu\nu}(\mathbf{q}, y, y')[1 - f(y')]}{y - i\omega_p} \right\}$$

where

$$J_{\mu\nu}(\mathbf{q}, y, y') = \frac{1}{v}\sum_{\mathbf{k}} k_\mu J_\nu(\mathbf{k}, \mathbf{k} + \mathbf{q}, y, y')$$

$$K_{\mu\nu}(\mathbf{q}, y, y') = \frac{1}{v}\sum_{\mathbf{k}} k_\mu K_\nu(\mathbf{k}, \mathbf{k} + \mathbf{q}, y, y')$$

It is clear that the analytic continuation to $F_{\mu\nu}(\mathbf{q}, \omega)$ is obtained by replacing $i\omega_p$ by $\omega_p$. We shall not rewrite the new formulas explicitly, but shall imagine $i\omega_p$ to be continuous. We then have to decide under what conditions we can let $\mathbf{q}$ and $\omega_p$ tend to zero in any order and obtain a unique result. If $J_{\mu\nu}(\mathbf{q}, y, y')$ is nonsingular as $\mathbf{q}$ tends to zero and $y$ tends to $y'$, and if $K_{\mu\nu}(\mathbf{q}, y, y')$ is nonsingular as $\mathbf{q}$ tends to zero and $y$ tends to zero, then the result of the limiting procedures is certainly independent of the order in which the limits are taken. In fact, the only singularities which make the result not unique are

$$\lim_{q\to0} J_{\mu\nu}(\mathbf{q}, y, y') \sim \gamma(y')\delta(y - y') \text{ or } \beta(y - y')^{-\alpha} \qquad \alpha \geqslant 1$$

$$\lim_{q\to0} K_{\mu\nu}(\mathbf{q}, y, y') = \gamma(y')\delta(y) \text{ or } \beta y^{-\alpha} \qquad \alpha \geqslant 1$$

(25)

But one can show directly from equations (23) and (24) that as $y'$ tends to $y$, $(y' - y)\lim_{q\to0} J_\mu(\mathbf{q}, y, y')$ tends to zero and also that as

$y$ tends to zero $y \lim_{q \to 0} K_\mu(\mathbf{q}, y, y')$ tends to zero. Hence, the only possible singularities are the delta-function ones. The necessary and sufficient conditions for the order of the limits to be unimportant is then that as $\mathbf{q}$ tends to zero, (a) $J_{\mu\nu}(\mathbf{q}, y, y')$ should not behave like $\gamma(y)\delta(y - y')$ when $y'$ approaches $y$, and (b) $K_{\mu\nu}(\mathbf{q}, y, y')$ should not behave like $\gamma(y')\delta(y)$ when $y$ approaches zero.

## 5. SPECIAL CASES

In the case of free particles the vertex part is as shown in Fig. 1.

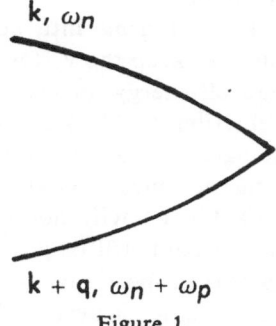

$$\mathbf{k}, \omega_n$$

$$\mathbf{k} + \mathbf{q}, \omega_n + \omega_p$$

Figure 1.

One then has

$$\Gamma_{\alpha\beta}(\mathbf{k}, \mathbf{k} + \mathbf{q}, \omega_n, \omega_n + \omega_p) = \frac{(2\mathbf{k} + \mathbf{q})}{m} G_\alpha(\mathbf{k}, \omega_n) G_\beta(\mathbf{k} + \mathbf{q}, \omega_n + \omega_p)$$

(25)

where $G(\mathbf{k}, \omega_n)$ is the single particle Green's function,

$$G(\mathbf{k}, \omega_n) = \frac{1}{i\omega_n + \epsilon_k}$$

and $\epsilon_k$ is the single particle energy. In this case the spectral function $K_\mu$ is zero and $J_\mu$ is a product of spectral functions for the individual particles, i.e.,

$$J_\mu(\mathbf{k}, \mathbf{k} + \mathbf{q}, y, y') = \frac{e}{m}(2k_\mu + q_\mu)\delta(y - \epsilon_k)\delta(y' - \epsilon_{k+q})$$

and

$$J_{\mu\nu}(\mathbf{q}, y, y') = \frac{e}{m}\sum_k k_\nu(2k_\mu + q_\mu)\delta(y - \epsilon_k)\delta(y' - \epsilon_{k+q})$$

As $\mathbf{q}$ tends to zero, we evidently have $J_{\mu\nu}$ proportional to $\delta(y - y')$.

This behavior of $J_{\mu\nu}$ is responsible for the anomalous behavior of free particles.

In the case of the BCS theory where one has independent quasi-particles, the vertex part is still given by equation (25), but in the notation of Nambu,[13] the Green's functions are given by

$$G(\mathbf{k}, \omega_n) = \frac{1}{i\omega_n + \epsilon_k \tau_3 - \Delta\tau_1} \tag{26}$$

To find $J$ and $K$, one has to take a trace in the space of the matrices $\tau$. The spectral functions are still $\delta$-functions and $J_{\mu\nu}$ still has a part proportional to $\delta(y - y')$. Within this approximation, then, the two effects are not equivalent.

As soon as we take scattering into account, the situation is changed. If as a first step we keep the vertex funtion in the form of equation (25) but include self-energy corrections in the single particle Green's functions, the function $J(\mathbf{k}, \mathbf{k}', y)$ is still a product of spectral functions for the single particles. Since these spectral functions, in the presence of scattering, no longer contain $\delta$-functions but instead have a finite width, $J(\mathbf{k}, \mathbf{k}'y, y')$ will not contain products of $\delta$-functions. In principle, it would still be possible for $J(\mathbf{q}, y, y')$, as $\mathbf{q}$ tends to zero, to have a part proportional to $\delta(y - y')$ coming from the summation of $J(\mathbf{k}, \mathbf{k} + \mathbf{q}, y, y')$ over $\mathbf{k}$. This could arise from oscillatory behavior of $J(\mathbf{k}, \mathbf{k} + \mathbf{q}, y, y')$ for large $\mathbf{k}$ or from finite singularities in this function for $\mathbf{k}$. In the cases that have been examined in detail, namely, impurity scattering and phonon scattering within the approximation that the conductivity is usually calculated, no such possibilities arise. Hence, one concludes that, within this approximation, $J(\mathbf{q}, y, y')$ has no singularity $\delta(y - y')$ in it.

This approximation, however, is not entirely consistent. If one includes self energy corrections one must also introduce corresponding corrections to the vertex part. To the second order in the scattering process one must sum diagrams of the type shown in Fig. 2. There the dashed line represents the scatterer, phonon or impurity. Now it can be shown by detailed inspection that each term in this sum can give rise to no singularity $\delta(y - y')$ in $J$ or $\delta(y)$ in $K(q)$. One must, however, be careful to ensure that the total sum gives rise to no such singularity. This question can be settled only by examining the integral equations for the sum. We need consider only the equation with $\mathbf{q}$ zero, and this has the form

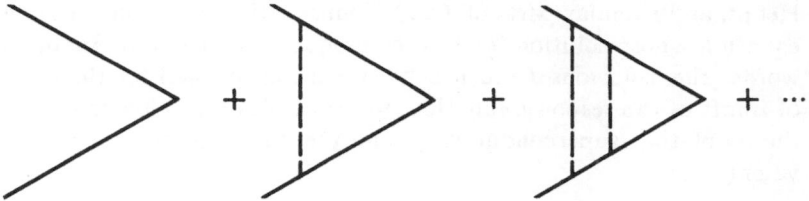

Figure 2.

$$\Gamma_\mu(\mathbf{k}, \mathbf{k}, \omega_n, \omega_n + \omega_p) = (k_\mu + k_\mu)\, G(\mathbf{k}, \omega_n + \omega_p)\, G(\mathbf{k}, \omega_n)$$

$$+ \frac{\beta}{v} \sum_{Q, \omega_l} G(k, \omega_n + \omega_p)$$

$$\times \tau_3 \Gamma_\mu(\mathbf{k} - Q, \mathbf{k} + Q, \omega_n - \omega_l, \omega_n - \omega_l + \omega_p)$$

$$\times \tau_3 G(k, \omega_n) D(Q, \omega_l) \tag{27}$$

where, for impurity scattering,

$$D(Q, \omega_l) = |V(Q)|^2$$

and for phonon scattering,

$$D(Q, \omega_l) = \frac{g^2 (\hbar \omega_Q)^2}{\omega_l^2 + (\hbar \omega_Q)^2}$$

The case of impurity scattering can be solved explicitly and it is well known that there is no singularity in the kernel.

For phonon scattering we have to examine the integral equation itself. We note that first of all that since $\Gamma$ gives the response of the system to an external field it should be unique. Further,† the functions $J$ and $K$ are determined uniquely by $\Gamma$ and, hence, by equation (27). If one substitutes the spectral form (22) in the integral on the right-hand side of equation (27), one finds that if $J$ and $K$ are nonsingular and possess no parts of the type in equation (25), then, since the integral over $\mathbf{q}$ is finite, the resulting integral is of the same spectral form, with new spectral functions $J'$ and $K'$ which are also nonsingular. Hence, the integration does not produce singularities of the unwanted type. However, as we have seen, the inhomogeneous term does not possess such singularities either.

---

†For further details and proofs of these statements the reader is referred to the original paper.[9]

Hence, any singular parts of $J$ and $K$ must satisfy the homogeneous equation whose solution for $\Gamma$ to be unique must be zero. In other words, the solutions $J$ and $K$ behave sufficiently well for the order of limits of the response function to be irrelevant. This completes the proof that superconductivity and the Meissner effect are equivalent.

## REFERENCES

1. J. Bardeen, L. N. Cooper, and Shrieffer J. R., *Phys. Rev.* **108**: 1175 (1957).
2. L. D. Landau, *J. Phys.* (*U. S. S. R.*) **5**: 71 (1941).
3. K. Maki, *Physics* **1**: 21, 127 (1964).
4. P. K. De Genues, "Superconductivity of Metals and Alloys," W. A. Benjamin Inc, New York, 1966.
5. A. A. Abrikosov, and Gorkov, L. P. *JETP* (*USSR*) **39**: 1781 (1960); Translation, *Soviet Phys.*, JETP **12**: 1213 (1961).
6. V. Ambegaokar, private conversation.
7. C. N. Yang, *Rev. Mod. Phys.* **34**: 694 (1962).
8. A. A. Abrikosov and L. P. Gorkov, *JETP* (*USSR*) **36**: 319 (1959); Translation, *Soviet Phys. JETP* **9**: 220 (1959).
9. W. A. B. Evans and G. Rickayzen, *Ann. Phys.* **33**: 275 (1965).
10. G. Wentzel, *Proc. Nat. Acad. Sci.* (*USA*) **49**: 679 (1963).
11. A. B. Pippard, *Proc. Roy. Soc.* (*Lond.*) **A216**: 547 (1963).
12. R. Kubo, *J. Phys. Soc. Japan* **12**: (1957).
13. Y. Nambu, *Phys. Rev.* **117**: 648 (1960).
14. L. Pičman, Nuklearni Insitut "Jožef Stefan," Ljubljana, preprint.

# Singularities of Multiple Scattering Processes

## F. PHAM

C.E.N.
Saclay, France

*and*

C.E.R.N
Geneva, Switzerland

---

## 1. INTRODUCTION

It will be useful to start this general study of Landau singularities by reviewing the simplest among them—the most important ones for many practical purposes:

1.   The *one-particle exchange graph* (Fig. 1) is currently interpreted as representing the "exchange" of a "virtual" particle; it contributes a pole for an *unphysical* value $t = m^2$ of the momentum transfer squared: $t \equiv (p_1 - p_3)^2$ ($m$ is the mass of the exchanged particle).†

2.   The *N-particle threshold* (or *normal threshold*) *graph* (Fig. 2) contributes a square root ($N$ even) or logarithmic ($N$ odd) branch point, at the value $s = (m_1 + m_2 + m_3 + \cdots + m_N)^2$ of the total energy squared: $s \equiv (p_1 + p_2)^2$. Thus this singularity occurs for a *physical* value of the total energy, corresponding to the "opening of an $N$-particle channel" ($m_1, m_2, \cdots, m_N$, are the masses of these $N$-particles).

It is generally believed that the normal thresholds are the only singularities which can occur in two-body scattering amplitudes for

---

†Throughout these notes, the letter $p$ will represent a four-vector $p = (p^{(0)}, p^{(1)}, p^{(2)}, p^{(3)})$ and $p \cdot p'$ will be the scalar product taken with the Lorentz metric:

$$p \cdot p' = p^{(0)} p'^{(0)} - p^{(1)} p'^{(1)} - p^{(2)} p'^{(2)} - p^{(3)} p'^{(3)}$$

$p^2$ is short for $p \cdot p$.

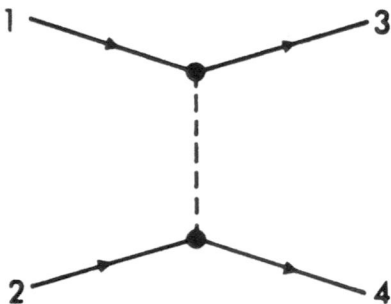

Fig. 1. One-particle exchange graph.

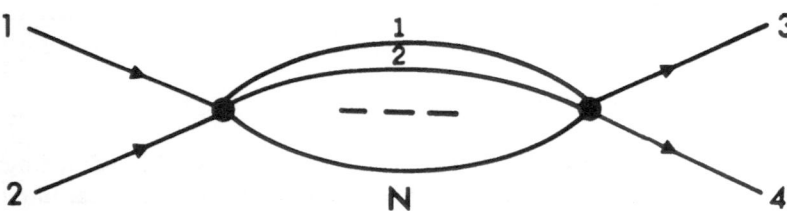

Fig. 2. $N$-particle threshold graph.

physical values of the momenta. In other words, the scattering amplitudes should be analytic functions of the momenta in the physical region between the successive normal thresholds. The importance of this belief in our way of thinking appears in the way resonances are usually defined, namely, as "poles in the second sheet near the physical region." This definition implicitly presupposes some analyticity in the "second sheet," at least in some neighborhood of the physical region. But such analyticity has never been proved, except (using the elastic unitarity) in the so-called "elastic region"†—a very unsatisfactory limitation since many important resonances lie outside this elastic region, above the three-particle thresholds. The difficulty with generalizing this result to the inelastic region is that one must use three-particle unitarity and, therefore, know something about the analytic properties of $2 \to 3$ amplitudes, but for the same reason, this implies knowing something about $3 \to 3$ amplitudes,

---

†And even there the "proof" is incomplete, as shown recently by A. Martin.

etc. Thus, one is led to the problem of simultaneously determining the analytic properties, in the physical region, of various many-particle scattering amplitudes using the unitarity relations which couple them together. This is of course a big program, and I am not even sure we have enough input information with out present knowledge of "physical sheet analyticity" as derived by the so-called "linear program" of axiomatic field theory. Nevertheless it may be interesting—if only to get an idea of what is awaiting us—to try to guess what kind of singularities must be present in the physical region of many-particle amplitudes. It has been known for some time that this singularity structure is much richer than in the two-particle case. For instance, $3 \to 3$ scattering amplitudes have polar singularities arising from one-particle exchange graphs, as represented in Fig. 3. Unlike that of Fig. 1, this singularity occurs in the physical region. One should notice that the "exchange" process is no longer virtual and can be observed, in a bubble chamber for instance, as a double scattering. But many other multiple scattering processes can be imagined, and it is tempting to think that they will all contribute to singularities in the physical region. This view is sustained by a very nice argument of Coleman and Norton who recently noticed that the intuitive idea of a multiple scattering, i.e., of a number of scattering processes occurring in a definite succession, leads to a very simple interpretation of the "Landau equations," usually derived to locate the singularities of a "Feynman graph." The following example will be enough to convey their main idea: Let us consider three successive collisions occurring as in Fig. 4, each at a given point A, B, and C of space–time. If $\tau_i$ is the time interval between the creation and the

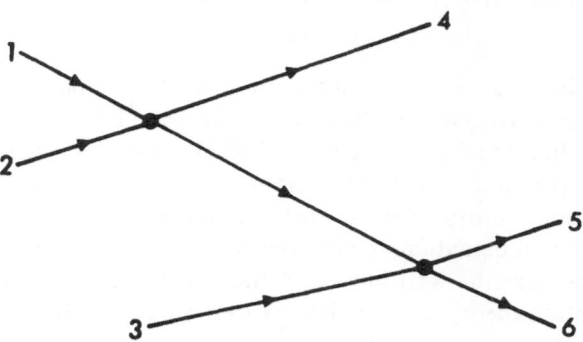

Fig. 3. Double scattering graph.

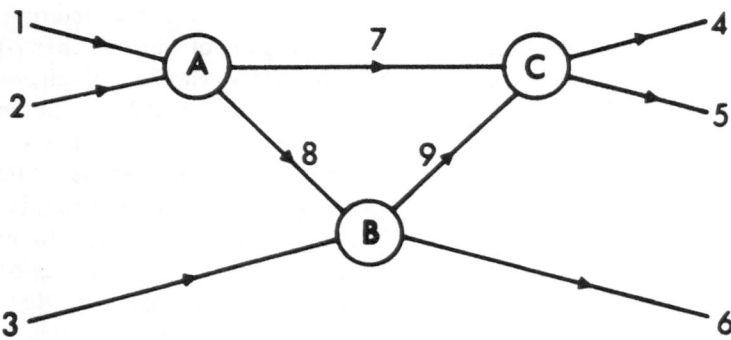

Fig. 4. Triple scattering graph.

destruction of the intermediate particle $i$ $(i = 7, 8, 9)$, measured in the rest system of that particle, then one can deduce the "four-velocities" $v_i$:

$$v_7 = \frac{\overline{AC}}{\tau_7} \qquad v_8 = \frac{\overline{AB}}{\tau_8} \qquad v_9 = \frac{\overline{BC}}{\tau_9}$$

hence the four-momenta

$$p_7 = \frac{\overline{AC}}{\alpha_7} \qquad p_8 = \frac{\overline{AB}}{\alpha_8} \qquad p_9 = \frac{\overline{BC}}{\alpha_9} \qquad \alpha_i = \frac{\tau_i}{m_i}$$

Conversely, given the four-momenta $p_i$ and the scalar parameters $\alpha_i$ the above relations determine uniquely, up to a common translation, the space–time location of the points $A$, $B$, and $C$, provided the following compatibility relation is satisfied:

$$\alpha_7 p_7 = \alpha_8 p_8 + \alpha_9 p_9$$

An analogous reasoning for the most general "multiple scattering graph" would give the compatibility relations

$$\sum_{i \in I} z(i)\alpha_i p_i = 0$$

where I is the set of all internal lines of the graph, and $z$ is any "cycle" constructed with these lines [$z(i) = +1$ or $-1$ if the cycle $z$ contains the line $i$ just once and with the "good" or "'wrong" orientation; $z(i) = 0$ if $z$ does not contain $i$]. If one writes, in addition, the above compatibility relations, the equations expressing the conservation of four-momentum at each vertex, and the mass shell conditions, one recognizes the well-known "Landau equations" of conventional perturbation theory. Notice that for the above intuitive interpretation to hold, the parameters $\alpha_i$ must be positive (a particle cannot be destroyed before it is created). Notice also that although the above interpretation is essentially classical, it is not inconsistent with quantum mechanics in the "macroscopic" limit, i.e., the limit where the successive scatterings are widely separated in space–time. In fact, although the uncertainty principle forbids us to consider strictly pointwise collisions, the extension of the collision region can be neglected if these regions are sufficiently far apart from one another (this idea will be stated more precisely in Section 4).

To summarize: We now have at our disposal a macroscopic space–time interpretation of the Landau equations for "multiple scattering graphs." In the next section we shall present a momentum–space interpretation of the same equations, showing that they determine "thresholds of multiple scattering channels" in complete analogy with the normal thresholds.

## 2. MULTIPLE SCATTERING THRESHOLDS

Let G be a multiple scattering graph. Henceforth, no meaning other than combinatorial is to be attached to the word "graph." The epithet "multiple scattering" refers to the physical interpretation, where the lines represent known particles with well-defined masses, whereas, the vertices represent collisions between these particles. Thus a multiple scattering graph is a mathematical device for representing the causal relationships between successive collisions. Mathematically, these causal relationships are expressed by the partial ordering on the set of vertices of the graph: a graph determines a partial ordering on its set of vertices if and only if it contains no "directed loop," i.e., if "no path following the direction of the arrows can lead us back to the original vertex."

This section is devoted to the following question: For which values of the external momenta is the multiple scattering process kinematically allowed? By this we mean the following: Given a system of four-vectors attached to the external lines of the graph, is it also possible to attach four-vectors to the internal lines preserving the law of momentum conservation at each vertex and the mass shell conditions for each line? (From now on, we shall always consider the positive energy conditions as part of the mass shell conditions; this amounts to throwing away one sheet of the two-sheeted mass hyperboloid).

Let us translate this question into geometrical language.

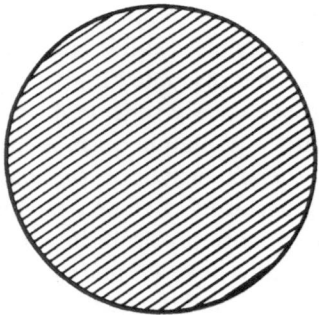

Fig. 5. Sphere projected on a plane (the plane of the picture).

To attach a four-vector to each line (external as well as internal) means to choose a point in $4|\mathbf{I}|$ — dimensional Euclidean space $\mathbb{R}^{4|\mathbf{I}|}$, where $|\mathbf{I}|$ is the total number of lines of $\mathbf{G}$ (we denote by $\mathbf{I}$ the set of all these lines). In $\mathbb{R}^{4|\mathbf{I}|}$, the variety† of those points which satisfy momentum conservation and mass shell conditions will be called the space of the graph and denoted by $\mathbf{M}$. Now if we denote by $G$ the "elementary scattering graph" deduced from $\mathbf{G}$ by contracting all the internal lines and by $M$ the space of $G$, we have an obvious "projection"

$$\rho: \mathbf{M} \longrightarrow M$$

(the mapping defined by forgetting the internal momenta), and the above question amounts to asking whether a given point $p \in M$ belongs to the image of $\mathbf{M}$ under the projection $\rho$.

In general, when one wants to study the projection of a manifold‡ it is useful to search for its "critical set," defined as the set of points in the manifold where the tangent space has a projection of smaller dimension than that of the goal. This is useful because the projection of the critical set—often called the apparent contour—"more or less" coincides with the boundary of the projection, as illustrated (more or less!) by Figs. 5, 6, 7, 8, and 9. In Figs. 5 and 6,

---

†Strictly speaking we should not call this set a "variety" since it is given not only by polynomial equations but also by inequalities (the positive energy conditions). Nevertheless, it is a connected component of a variety.

‡A manifold is a variety without singular points, i.e., at every point the tangent space can be defined. We have to suppose that the space of the graph is a manifold because I am not aware of any satisfactory definition of the critical set for varieties with singular points. Proposition 2 below shows that this restriction is not too drastic.

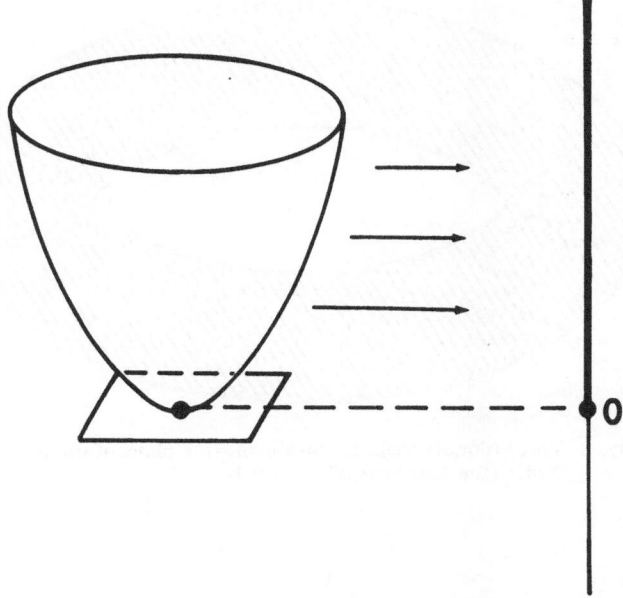

Fig. 6. Paraboloid projected on a line.

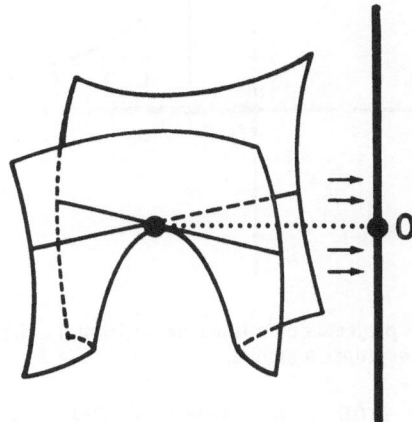

Fig. 7. Hyperbolic paraboloid (saddle) projected on a line.

the apparent contour coincides with the boundary of the projection. In Figs. 7 and 8, it is greater. It can be proved that the apparent contour always contains the boundary of the projection, provided

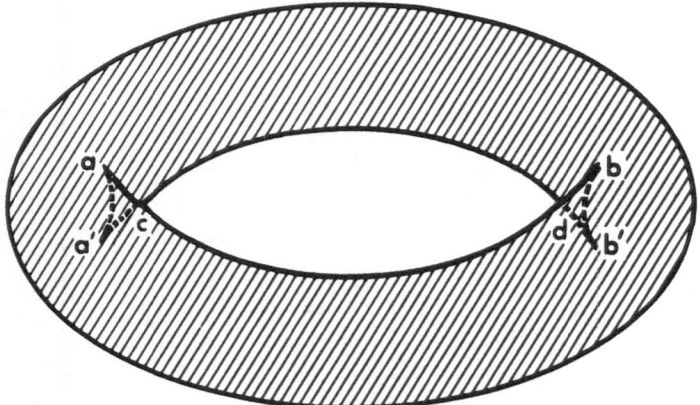

Fig. 8. Torus (donut) projected on a plane (the plane of the picture).
Notice the four "cusps" a, a′, b, b′.

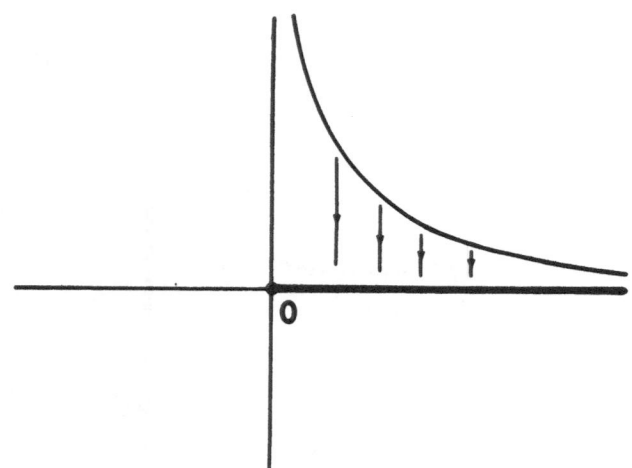

Fig. 9. Curve projected on a line (the horizontal axis): an example
of a *nonproper* mapping.

the mapping is proper, i.e., provided the reciprocal image of
every compact set is compact. Figure 9 shows an example of a nonproper mapping, where the projection has a non-empty boundary
although there is no critical point (loosely speaking, one could say
that "the critical point is at infinity"). Intuitively, our task will be

to discard the "bad" cases symbolized in Figs. 7, 8, and 9, so that the apparent contour will coincide with the boundary of the projection, therefore deserving to be called "threshold of the multiple scattering process"—a physical interpretation which will appear the more remarkable when we discover (*Proposition 3*) that the apparent contour is given by the Landau equations. Let us prove first† the following:

*Proposition 1: The mapping $\rho$ is proper.* Obviously a closed subset of the mass hyperboloid is compact (i.e., bounded) if and only if the energy is bounded. Therefore we only have to prove that the boundedness of the energies for the external lines implies the boundedness of the energies for the internal lines. Let us choose a vertex $v$ which is maximal for the "causal" order relation of the graph, i.e., no internal line goes out of that vertex. Then the internal energies will appear with the same sign in the energy conservation law of that vertex, and since they are all positive, their boundedness will be ensured by the boundedness of the external energies. Having thus proved the boundedness for all the internal lines incident to the vertex $v$ we can start again with a vertex which is maximal among the remaining vertices, etc., until the whole graph is exhausted.

*Proposition 2: For "generic" values of the masses, the space of a graph is a manifold.* In fact, denoting by $\mathbf{E} \subset \mathbb{R}^{4|\mathbf{I}|}$ the linear subspace defined by the conservation of momentum at each vertex of the graph $\mathbf{G}$, and by $\mathbf{s}: \mathbf{E} \longrightarrow \mathbb{R}^{|\mathbf{I}|}$ the mapping which associates the numbers $\mathbf{s}_l = p_l^2$ to the system of momenta $(p_l)_{l \in \mathbf{I}}$, we can consider the space $\mathbf{M}$ as the reciprocal image of the point $\mathbf{m}^2 = (m_l^2)_{l \in \mathbf{I}}$ through the mapping $\mathbf{s}$ (or rather a connected component of this reciprocal image). Such a space will be a manifold (see Theorem 2 of Ref. 2) provided $\mathbf{m}^2$ does not belong to the apparent contour of $\mathbf{s}$, which is a "meager"‡ set. We shall call "generic" those values of $\mathbf{m}^2$ which do not belong to the apparent contour.

It is instructive to explicitly write what the critical set of the

---

†Among the five following propositions, the first and the last are quite simple to prove. To understand in full detail the proofs of the other three the reader will need some knowledge of differential topology, which is provided in Ref 2.

‡A meager set is one whose adherence is nowhere dense—it is a general fact that the apparent contour of a polynomial mapping is a "semialgebraic set" contained in a proper algebraic subset of the goal.

mapping s is. As is well known, a system of momenta $(p_l)_{l \in I}$ obeying the conservation law at each vertex admits of a unique development

$$p_l = \sum_k z_k(i) p_k$$

once chosen a basis $(z_k)$ of the group of cycles of the graph $G$ (the integer $z_k(i)$ is the coefficient of the line $i$ in the cycle $z_k$). Taking the system of momenta $p_k$ as independent coordinates on $E$, we can thus write the Jacobian matrix of the mapping

$$\left\| \frac{\partial s_l}{\partial p_k} = 2 z_k(i) p_l \right\|$$

Therefore a point $(p_l) \in E$ will be critical for $s$ if and only if there exist parameters $\alpha_l$, not all zero, such that

$$\sum_{l \in I} \alpha_l z(i) p_l = 0$$

for all cycles $z$ of the graph. Notice that the above conditions look like the Landau equations, except that they deal with all the cycles of the graph, and not only the internal cycles.

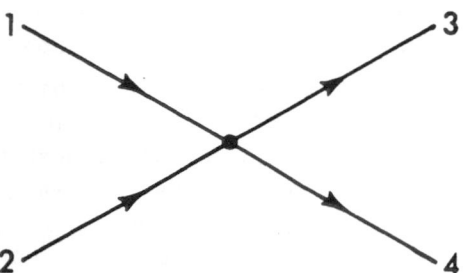

Fig. 10. Elementary scattering graph.

*Example:* On an "elementary graph" such as that of Fig. 10, the group of cycles can be generated by $1 - 2, 1 + 3, 1 + 4$, so that the above critical conditions read

$$\alpha_1 p_1 - \alpha_2 p_2 = 0 \qquad \alpha_1 p_1 + \alpha_3 p_3 = 0 \qquad \alpha_1 p_1 + \alpha_4 p_4 = 0$$

meaning that the four-vectors $p_l$ are all proportional to each other, but this is possible only if the masses bear the following relationship: $m_1 + m_2 = m_3 + m_4$. This example shows that nongeneric values of the masses often occur in physics, in particular for all "elementary graphs" describing elastic scattering processes. Nevertheless, owing to their "nongeneric" character (i.e., that the corresponding masses form a meager set), such cases can always (at least in principle) be studied by a limiting procedure, starting from generic values.

### Local Singularities of the Mapping $\rho$

Suppose the values of the masses are generic, so that both spaces M and $M$ are manifolds. We want to study the singularities of the mapping $\rho: M \longrightarrow M$. In effect, taking advantage of Section 2 of Ref. 2, we shall study instead the singularities of the mapping $\sigma$ shown on the following commutative diagram:

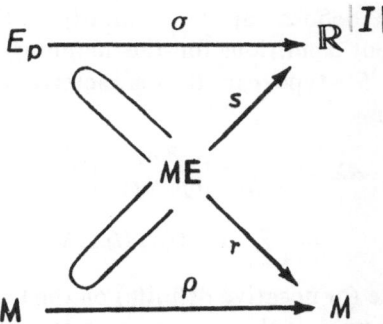

where $M\,E$ is the subspace of $E$ defined by the mass shell conditions for the external lines only, $r$ is the canonical projection onto the "external space" $M$, and $E_p$ is the linear subspace, "fiber of $r$ above the point $p \in M$":

$$E_p = r^{-1}(p)$$

$s$ and $\sigma$ are the mappings which associate to the point $\mathbf{p} = (p_i)_{i \in I}$ the system of numbers

$$(s_i = \sigma_i = p_i^2)_{i \in I}$$

where I is the set of internal lines of the graph. One can define a system of independent coordinates on $E_p$ by choosing a basis $z_k$ of the group of internal cycles of the graph, and associating to each member of this basis a four-vector $p_k$. The momentum of the internal line $i$ will be given by

$$p_i = \sum_k z_k(i)p_k + \text{ext. mom.}$$

where ext. mom. is some well-defined combination of external momenta, which we need not write explicitly. In these coordinates, the tangent mapping to $\sigma$ is defined by the matrix

$$\frac{\partial \sigma_i}{\partial p_k} = 2z_k(i)p_i$$

Therefore the point $\mathbf{p}$ will be critical if and only if it satisfies the Landau equations

$$\sum_{i \in I} \alpha_i z_k(i) p_i = 0 \qquad \forall z_k$$

for some not-all-vanishing parameters $\alpha_i$. The *corank* (at the goal) of the critical point will be the dimension of the vector space of the $\alpha$'s satisfying the above equation. Supposing the corank is 1, i.e., the $\alpha$'s are well defined up to a multiplicative factor, let us search for sufficient conditions for the mapping to have type $S_+^1$. Recall that the $S_+^1$ type can be characterized by the fact that the quadratic form

$$AX = \sum_{\substack{i,k,k' \\ \lambda,\mu}} \alpha_i \frac{\partial_3 \sigma_i}{\partial p_k^{(\lambda)} \partial p_k^{(\mu)}} X_k^{(\lambda)} X_{k'}^{(\mu)}$$

$$= \sum_{i,k,k'} \alpha_i z_k(i) z_{k'}(i) X_k X_{k'}$$

is positive definite (or negative definite) on the kernel of the tangent map to $\sigma$.† The kernel of this tangent map consists of those $X$ which satisfy

$$\sum_{k,\lambda} \frac{\partial \sigma_i}{\partial p_k^{(\lambda)}} X_k^{(\lambda)} \equiv \sum_k z_k(i) p_i \cdot X_k = 0 \qquad \forall i \in I$$

If we set $Y_i = \sum_k z_k(i) X_k$, we can rewite $A(X)$ as

$$A(X) = \sum_{i \in I} \alpha_i Y_i^2$$

and the "kernel condition" as

$$p_i \cdot Y_i = 0 \qquad \forall i$$

Since the $p_i$'s are timelike vectors, this kernel condition tells us that the $Y_i$'s are spacelike, i.e., $Y_i^2 \leqslant 0$, with the equality holding only if $Y_i = 0$. Therefore $A(X)$ will be negative semidefinite if the $\alpha_i$'s are $\geqslant 0$. More careful analysis is required before we come to any conclusions about the definiteness. Suppose that some $X$ exists for

---

†The $X_k$'s are, like the $p_k$'s, four-vectors. By $X_k^{(\lambda)}$ we mean the components of these four-vectors ($\lambda = 0, 1, 2, 3$).

which $A(X) = 0$. Then each term $\alpha_i Y_i^2$ must be zero, so that the four-vector $Y_i$ must be zero for every $i$ such that $\alpha_i \neq 0$. But notice that these four-vectors, by their very definition in terms of the $X_k$'s, must satisfy the conservation laws of the "internal graph," the graph obtained by removing all the external lines. Obviously the conservation laws are not affected by also removing those internal lines for which $Y_i = 0$. We are thus left with a graph for which all $\alpha_i$'s are zero, and $Y_i \neq 0$. Such a graph cannot be a tree (i.e., it must have cycles), for otherwise its conservation laws would admit only the zero solution $Y_i = 0 \ \forall \ i$. We thus conclude that nondefiniteness is possible only if there exists an internal cycle on which all $\alpha_i$'s vanish.

The following two propositions summarize the conclusions of this discussion:

*Proposition 3: A point* $\mathbf{p} \in \mathbf{M}$ *is critical for* $\rho$ *if and only if it satisfies the Landau equations*

$$\sum_{i \in I} \alpha_i z(i) p_i = 0 \qquad \forall z$$

*for some not-all-vanishing parameters* $\alpha_i$. *The corank of the critical point is the dimension of the vector space of the* $\alpha$'s *satisfying the above equations.*

*Definition 1: A critical point* $\mathbf{p} \in \mathbf{M}$ *is called a principal critical point if its corank is 1, the parameters* $\alpha_i$ *are non-negative, and every internal cycle has a line* $i$ *with* $\alpha_i$ *strictly positive.*

*Proposition 4: Near a principal critical point the mapping* $\rho$ *is of type* $S^1_+$, *i.e., one can choose local coordinates* $x_1, x_2, \cdots, x_m$ *in* $\mathbf{M}$ *and local coordinates* $y_1, y_2, \cdots, y_n$ *in* $M$ *such that* $\rho$ *read:*†

$$y_1 = x_1$$
$$y_2 = x_2$$
$$\cdots \cdots$$

---

†Here we suppose $m = \dim \mathbf{M} \geqslant n = \dim M$. But if $m = n - 1$, all points of $\mathbf{M}$ are "critical" (the "apparent contour" is simply the projection of $\mathbf{M}$), and those of corank 1 correspond to the local model

$$y_1 = x_1$$
$$y_2 = x_2$$
$$\cdots \cdots$$
$$y_{n-1} = x_{n-1}$$
$$y_n = 0$$

which expresses that $\rho$ is an "immersion" (an important example is the one-particle exchange graph). The cases $m < n - 1$ are excluded by Definition 1, since their corank cannot be 1.

$$y_{n-1} = x_{n-1}$$
$$y_n = x_n^2 + x_{n+1}^2 + \cdots + x_m^2$$

Let us recall some important features of $S_+^1$ mappings.

The critical set is a manifold—given by $x_n = x_{n+1} = \cdots = x_m = 0$ in the above coordinates—(locally) isomorphic to the apparent contour, which is itself a manifold of codimension 1—given by $y_n = 0$ in the above coordinates. The space M is locally projected all on the same side of the apparent contour (the side $y_n \geqslant 0$), and there its "fiber" $\rho^{-1}(p)$ is a "vanishing sphere" (the sphere $x_n^2 + x_{n+1}^2 + \cdots + x_m^2 = y_n$ which shrinks to zero when $y_n$ goes to zero).

*Physical examples:* Applying the above result to the "two-particle threshold graph," we find that the "two-particle phase space" is a sphere, which shrinks to zero at the threshold. Not so well known perhaps is the analogous result for $N$-particle thresholds: The $N$-particle phase space is homeomorphic to a $3N - 4$ dimensional sphere. Proposition 4 provides a local generalization of these results to arbitrary "multiple scattering graphs," as far as "principal critical points" are concerned.

*Mathematical examples:* In Figs. 5 and 6, the projection is $S_+^1$ at all critical points. In Fig. 8 also, except at the "cusps" a, b, a′, b′. In Fig. 7, the critical point is $S^1$, but not $S_+^1$ (the quadratic form is nondegenerate, but not positive definite).

## What About the Global Structure?

Despite Proposition 4, which tells us that the projection under study is locally very well behaved, some strange things could happen in its *global* behavior. For instance, branches of apparent contour, coming from different critical points, could cross each other, as at the point $c$ (or $d$) of Fig. 8. Near such points there would be pieces of apparent contour which would not belong to the boundary of the *global* projection (although, according to Proposition 4, they would still belong to the boundary of the projection restricted to a small neighborhood of the critical point in M). Proposition 5 below, proved by quite elementary means, shows that such strange things cannot happen here, at least provided that one forgets about the mass shell conditions for the external lines. This means that instead of studying the projection M $\longrightarrow$ $M$ we study the projection ME $\longrightarrow$ $E$, where ME $\subset$ E is the variety defined by the mass shell conditions for the internal lines only (it is clear that the critical points of this new projection are given by the same Landau equations as previously).

*Definition 2: A critical point is called relevant if the Landau equations admit a solution with non-negative $\alpha_i$'s.*

*Proposition 5: Through every point $p^c$, projection of a relevant critical point $\mathbf{p}^c$, there passes a hyperplane which supports the projection of ME. It follows in particular that such a point $\mathbf{p}^c$ is a boundary point for the projection of ME (it also follows that the projection of ME has some convexity properties).*

Indeed, let $(\alpha_i)$ be a system of non-negative solutions of the Landau equations at the point $p^c$. On the linear space $\mathbf{E}$, consider the linear function

$$t(\mathbf{p}) = \sum_{i \in I} \alpha_i p_i^c \cdot (p_i - p_i^c)$$

It vanishes at the critical point, and it is non-negative on ME; indeed, none of the terms $p_i^c \cdot (p_i - p_i^c)$ can be negative when both four-vectors $p_i$, $p_i^c$ belong to the same mass shell with positive energies. Therefore $\{t(\mathbf{p}) = 0\}$ is a supporting hyperplane for ME in $\mathbf{E}$. Now it immediately follows from the critical character of the point $\mathbf{p}^c$ that this hyperplane is the reciprocal image of a hyperplane in $E$, i.e., the function $t(\mathbf{p})$ depends on the external momenta only. This can be deduced directly from the Landau equations, substituting in $t(\mathbf{p})$ the expressions

$$p_i = \sum_k z_k(i)p_k + \text{ext. mom.}$$

already mentioned above. Thus $\{t(\mathbf{p}) = 0\}$ defines a hyperplane in $E$, which supports the projection of ME.

In conclusion let us give some extra definitions which will be useful in the sequel.

*Definition 3: A critical point $\mathbf{p} \in \mathbf{M}$ is called a leading critical point if its corank is 1 and the parameters $\alpha_i$ are strictly positive.* Notice the logical implications "leading" (*Definition. 3*) $\Rightarrow$ "principal" (*Definition. 1*) $\Rightarrow$ "relevant" (*Definition. 2*).

*Definition 4: A "Landau point" $p \in M$ is the image of a critical point.* A Landau point is called "leading," respectively, "principal," respectively, "relevant" if it is the image of a "leading," respectively, "principal," respectively, "relevant" critical point.

*Remark:* In the literature, one uses to define Landau points by the Landau equations with either $p_i^2 = m_i^2$ or $\alpha_i = 0$, and leading Landau points as those where all the $\alpha_i$'s differ from zero. Our

definition of Landau points, derived from Proposition 3, stands intermediate between these two; the $\alpha_i$'s need not be all different from
zero but all the lines must be on the mass shell.


## 3. SINGULAR STRUCTURE OF THE $S$-MATRIX

In this section we state what kind of analyticity may be expected for the $S$-matrix in the physical region. The conjectures stated
below are more or less classical, but only thanks to the results of the
previous section can they be stated in such a precise fashion. In my
opinion, the plausibility of these conjectures is very strongly supported, a) by the physical arguments already mentioned at the end of
Section 1, and developed further in Section 4, b) by the mathematical
consistency arguments of Sections 5 and 6. Also these conjectures
can be proven in perturbation theory (see the introduction of Ref. 1).
We first recall some well-known facts, just to fix the notations.

If $\varphi$ and $\psi$ are free-particle states, the $S$-matrix element
$\langle\psi|S|\varphi\rangle$ is the probability amplitude of getting $\psi$ as an outgoing
state if the ingoing state was $\varphi$. It is convenient to intoduce the
"integral kernel," $\langle p^{I_2}|S|p^{I_1}\rangle$ ($S$-matrix element in momentum space")
defined by the symbolic expression

$$\langle\psi|S|\varphi\rangle = \int_I \langle p^{I_2}|S|p^{I_1}\rangle\,\psi(p^{I_2})\phi(p^{I_1})$$

where $I_1$ (respectively, $I_2$) is the set of ingoing (respectively, outgoing)
particles, $p^{I_1}$ (respectively, $p^{I_2}$) is the system of four-momenta of
these particles, and the $\int_I$ is the integration on the mass shell of all
these particles ($I = I_1 \cup I_2$), with the Lorentz-invariant measure
$\prod_{i\in I} d^4p_i\delta(p_i^2 - m_i^2)$. The above "integral kernel" is a distribution,
which can be factorized as

$$\langle p^{I_2}|S|p^{I_1}\rangle = \delta^4(p_{I_1} - p_{I_2})\langle p^{I_2}|S|p^{I_1}\rangle$$

denoting by $p_{I_1}$ (respectively, $p_{I_2}$) the total ingoing (respectively,
outgoing) momentum. $\langle p^{I_2}|S|p^{I_1}\rangle$ can be considered as a distribution
on the manifold† $M$ (see Section 3) defined by the mass shell conditions and the total momentum conservation conditions. It is conven-

---

†Difficulties occur when $M$ is not manifold (see Hepp). We shall not
bother about them.

ient to introduce "truncated amplitudes" $\langle p^{I_2}|S|p^{I_1}\rangle_T$; they can be defined iteratively by the following decompositions of the ordinary amplitudes:

$$\langle p^{I_2}|S|p^{I_1}\rangle = \sum_{\mathcal{K}} \prod_{K \in \mathcal{K}} \langle p^{I_2 \cap K}|S|p^{I_1 \cap K}\rangle_T$$

where the summation runs over all partitions $K$ of $I$. Each truncated amplitude can also be factorized through a momentum conservation $\delta$ -function:

$$\langle p^{I_2}|S|p^{I_1}\rangle_T = \delta^4(p_{I_1} - p_{I_2}) \langle p^{I_2}|S|p^{I_1}\rangle_T$$

These distributions $\langle p^{I_2}|S|p^{I_1}\rangle_T$ are the ones which are expected to have some analyticity properties (on the manifold $M$). The singular points are expected to be the "Landau points" of all multiple scattering graphs $\mathbf{G}$ which can be associated to the elementary scattering $I_1 \longrightarrow I_2$, i.e., all connected graphs having $I_1$ and $I_2$ as ingoing and outgoing sets of particles.

*Conjecture A: The distribution $\langle p^{I_2}|S|p^{I_1}\rangle_T$ is equal to an analytic function everywhere on M, except at the relevant Landau points of all multiple-scattering graphs which can be associated to the elementary process $I_1 \longrightarrow I_2$.*

Now since Landau points form sets of codimension 1(in general), they separate the real manifold $M$ into several disconnected pieces. In the following conjecture, we show how the Landau loci can be crossed through, allowing small imaginary distortions, in order to connect together the analytic functions defined in each piece.

*Definition 5: A Landau point is called smooth if all Landau points of all possible graphs form, in the neighborhood of this point, an analytic submanifold L of codimension 1.* An example of a smooth Landau point is, according to Section 2, any leading Landau point of a graph $\mathbf{G}$, provided no other graph adds extra Landau points in the neighborhood.

Near a smooth Landau point, $L$ can be given by equating to zero a real analytic function $l$ with $dl \neq 0$. If the point is relevant, we have seen (Proposition 5) that L can be considered as the "threshold" of a multiple scattering process. We shall always agree to choose $l$ so that $l > 0$ above the threshold.†

---

†In case several graphs give the same $L$, we have to suppose that they give the same definition of "above the threshold." A very general situation where two graphs can give the same $L$ is when one of them is a contraction of the other, but then one easily sees that their spaces project on the same side on $L$, so that there is no trouble.

*Conjecture B: In a sufficiently small neighborhood U of any smooth relevant Landau point, the distribution $\langle p^{I_2}|S|p^{I_1}\rangle$ is the boundary value of an analytic function (possibly multivalued) defined in $U^c - L^c$, where $U^c$ is a small complex extension of the real neighborhood U, and $L^c$ is the manifold deduced from L by complexification (i.e., $L^c$ is given by $l^c = 0$, where $l^c$ is the complex analytic function corresponding to l). This boundary value has to be taken through the complex "half neighborhood" Im $l^c > 0$.*

This choice of the upper "half neighborhood" Im $l^c > 0$ rather than the "lower half," is the essential part of the conjecture. It will be justified in Section 4.

*Absorptive Parts.* The above conjecture has defined some determination of an analytic function in the "upper half neighborhood" $U^c \cap \{$Im $l^c > 0\}$, through this determination was the physical amplitude obtained. Let us now consider, in the "lower half neighborhood" $U^c \cap \{$Im $l^c < 0\}$, the analytic function which coincides with the previous one for $l^c$ real under the threshold. Its boundary value will be denoted by $\langle p^{I_2}|S|p^{I_1}\rangle_T^L$. The "absorptive part along" L is the distribution

$$\langle p^{I_2}|A|p^{I_1}\rangle^L = \langle p^{I_2}|S|p^{I_1}\rangle_T - \langle p^{I_2}|S|p^{I_1}\rangle_T^L$$

defined in the real neighborhood U. We shall also use the notation

$$\langle p^{I_2}|\mathbf{A}|p^{I_1}\rangle^L = \delta^4(p_{I_1} - p_{I_2})\langle p^{I_2}|A|p^{I_1}\rangle^L$$

*Conjecture C:* ("Generalized unitarity" or "Cutkosky's rule"). *Let L be the Landau manifold in the neighborhood of a smooth Landau point, leading Landau point of one graph* **G** *and of no other. If* **G** *has only "simple lines,"† the absorptive part along L is given by the expression*

$$\langle p^{I_2}|\mathbf{A}|p^{I_1}\rangle^L = \int_I \prod_v \langle \mathbf{S}_v\rangle_T$$

where the product runs over all vertices of **G**, $\langle \mathbf{S}_v\rangle_T$ is for the S-matrix element of the "elementary process" occurring at vertex $v$, $\int_I$ is short for the integration over the mass shell of all internal particles with

---

†If **G** has "multiple lines," i.e., bunches of lines with the same origin and the same extremity, the statement Conjecture C is not so simple : the modified statement can be found—and justified to some extent—in Chapter III of Ref. 1.

the Lorentz invariant measure $\prod_{i \in I} d^4 p_i \, \delta(p_i^2 - m_i^2)$ (I is the set of internal particles of G).

The above formula will eventually become more convenient under the following form, deduced from the above by "extracting the momentum-conservation $\delta$ function":

$$\langle p^{I_2} | A | p^{I_1} \rangle^L = \int_{\rho^{-1}(p)} \prod_v \langle S_v \rangle_\tau$$

this time the integration is on the "fiber" of the mapping $\rho : M \to M$ (see Section 2), with the obvious measure $\prod_k d^4 p_k \prod_i \delta (p_i^2 - m_i^2)$, where $p_k$ is the set of independent momenta associated to a basis of the group of internal cycles of the graph G (see again Section 2).


## 4. MULTIPLE SCATTERING REVEALED IN THE SPACE–TIME ASYMPTOTIC BEHAVIOR OF THE $S$-MATRIX

The purpose of this section is to study how the probability amplitude $\langle \psi | S | \varphi \rangle$ of a process $I_1 \to I_2$ behaves asymptotically when the wave packets of the various particles undergo suitable space–time translations, designed so as to "put them in the conditions of observation of some multiple-scattering process G. To every vertex $v$ of the graph G we shall associate a four-vector $a_v$, by which will be translated all the external particles incident to $v$. The choice of the system $a_v$ will be thus determined: Let $p^c \in M$ be a leading Landau point of the graph G (and of no other), supposedly representing the mean momenta of the external particles. According to section 2, this point can be associated to a unique system of internal momenta $p_i{}^c$, and of positive parameters $\alpha_i$ defined up to a common multiplicative factor. The Coleman–Norton analysis suggests that we should interpret this multiplicative factor $\tau$ as a sort of "time scale of the process," and consider the four vectors $a_i = \tau \alpha_i p_i^c$ as giving the space–time separations of the various collisions involved in the multiple scattering process. Thus, we shall determine the $a_v$ by the conditions

$$a_i = a_{v_i{}''} - a_{v_i{}'}$$

where $v_i{}'$ and $v_i{}''$ are the origin and the extremity of the line $i$ (this determines the $a_v$'s up to the addition of a common vector, irrelevant since the $S$-matrix is translation invariant). Under the chosen translations, the truncated $S$-matrix element

$$\langle\psi|S|\varphi\rangle_T = \int_I \langle\rho^I|S|p^{I_1}\rangle_T \psi(p^{I_1})\varphi(p^{I_1})$$

becomes

$$\langle S\rangle(\tau) \equiv \langle\psi_a|S|\varphi_a\rangle_T$$

$$= \int_I \langle p^{I_1}|S|p^{I_1}\rangle_T e^{i\sum_v p_v \cdot a_v} \psi(p^{I_1})\varphi(p^{I_1})$$

where $p_v$ is the "algebraic sum" of the external momenta incident to the vertex $v$ (plus sign for incoming momenta; minus sign for outgoing). It is also equal, thanks to momentum conservation, to the "algebraic sum" (with the opposite sign) of the internal momenta incident to $v$. We thus find by a simple calculation

$$\sum_v p_v \cdot a_v = -\sum_{i \in I} p_i \cdot a_i$$

so that the phase factor in the integral reads

$$-\sum_{i \in I} p_i \cdot a_i = -\tau \sum_{i \in I} \alpha_i p_i^c \cdot p_i = -\tau[t(p) + \sum_{i \in I} \alpha_i m_i^2]$$

where the linear function $t(p)$ defined in Section 2 (proof of proposition 5) appears. Recall that this linear function has been found to be the equation of the tangent hyperplane to the Landau manifold (outside the mass shell) at the point $p^c$. This factor $e^{-i\tau t(p)}$ under the integral makes the function $\langle S\rangle(\tau)$ look like a sort of Fourier transform. It is well known that the Fourier transform of a function analytic in the upper half plane $\{\text{Im } t > 0\}$ decreases quickly as $\tau$ goes to $-\infty$. Here Conjecture B tells us that the function $\langle p^{I_1}|S|p^{I_1}\rangle_T$ is analytic in a small "upper half neighborhood" $U^c \cap \{\text{Im } l^c > 0\}$. Of course it is not so easy to get at a conclusion since a) $t$ is not exactly $l$, b) the analyticity property does not hold in a half plane but only a half neighborhood, and c) the integrand also contains the wave function $\psi(p^{I_1})\varphi(p^{I_1})$, which has no analyticity property. Nevertheless the conclusion that $\langle S\rangle(\tau)$ decreases faster than any inverse power of $\tau$ when $\tau \to -\infty$ can be shown to hold true (see Appendix 0 of Ref. 1) provided $\psi(p^{I_1})\varphi(p^{I_1})$ is an infinitely differentiable function with compact support contained in the neighborhood U.† When $\tau \to +\infty$, an analogous conclusion holds for the

---

†This is a rather natural hypothesis, since a) *smooth* wave packets are the most commonly used in physics, and b) in order to observe the multiple scattering process G and no other, one must work with wave packets whose support meets only the Landau set of G.

function $\langle S \rangle^L(\tau)$ defined by replacing the distribution $\langle p^{I_2}|\mathbf{S}|p^{I_1}\rangle_T$ by the distribution $\langle p^{I_2}|\mathbf{S}|p^{I_1}\rangle_T^L$ of Section 3, which is a boundary value from the opposite half neighborhood. But now Conjecture C gives for the function $\langle A \rangle^L(\tau) = \langle S \rangle(\tau) - \langle S \rangle^L(\tau)$ a very suggestive expression, allowing us to enunciate the following:

*Proposition 6: Modulo a fastly decreasing function, $\langle S \rangle(\tau)$ behaves when $\tau \to + \infty$ like the function*

$$\langle A \rangle^L(\tau) = \int_{I_1, I_2} \left(\prod_v \langle \mathbf{S}_v \rangle_T e^{-i \sum_{i \in I} p_i \cdot a_i} \bar{\psi}(p^{I_2}) \varphi(p^{I_1})\right)$$

This expression contains exactly the $S$-matrix factorization which would be suggested by interpreting the vertices of the graph as independent processes. The phase shift $p_i \cdot a_i$ comes from the propagation of the "free" particle $i$ between the point where it is created and the point where it is destroyed. This phase shift "damps" the amplitude of the process; in effect one can show that $\langle A \rangle(\tau)$ goes to zero as an inverse power of $\sqrt{\tau}$.†

Notice that the above discussion gives a physical reason why Conjecture C should not take exactly the same form for a graph with multiple lines. In fact, intermediate particles which "travel together" cannot be considered as free particles, i.e., they cannot behave like asymptotic *out* (respectively, *in*) particles at the vertex where they are created (respectively, destroyed). Therefore one is not justified in associating the ordinary $S$-matrix element $\langle \mathbf{S}_v \rangle_T$ to such a vertex.

## 5. SINGULARITIES OF ABSORPTION INTEGRALS

Conjecture C in Section 3 has given an integral representation for absorptive parts, the so-called "generalized unitarity" or "Cutkosky's rule." Since the integrand is just a product of $S$-matrix elements, its analytic properties are known by the Conjecture of Section 3, so that one should be able to deduce the analytic properties of the integral. This process of deducing the analytic prop-

---

†This is exactly the "spreading of wave packets" phenomenon, familiar from quantum mechanics. Mathematically one deals with the integral of an infinitely differentiable function multiplied by $e^{-i\tau t(p)}$. Evaluating this integral by the stationary phase method, one sees that it behaves asymptotically like $\tau^{-n/2}$, where $n$ is the rank of the quadratic form by which the Taylor development of the function $t|\mathbf{M}$ begins, at the point $\mathbf{p}^c$.

erties of "absorption integrals" from the analytic properties of
their integrand has played a prominent role in all the attempts
made so far of deducing Landau singularities without relying on
perturbation theory. This suggests that the "future proofs" of
Conjecture A, B, and C shall rely heavily on the study we are going
to make now.

For the sake of simplicity, we shall suppose here that all the
graphs we shall have to deal with are without multiple lines. We
have already given some hints as to why complications arise for
multiple lines. That these complications would not affect our conclusions is shown in Chapter III of Ref. 1.

*Definition 6: Absorption integrals.* Let **G** be a multiple scattering
graph, and $G$ be the graph of the associated elementary process. As
in Section 2, denote by $\rho: \mathbf{M} \to M$ the canonical mapping from the
space of **G** to the space of $G$. *The absorption integral of* **G** *at a point*
$p \in M$ *above the threshold of* **G** (*i.e.*, $p$ *belongs to the projection of*
**M**) *is defined as the integral*

$$A_G(p) = \int_{\rho^{-1}(p)} \prod_v \langle S_v \rangle_T$$

where the notation has the same meaning as in Conjecture C.

*Remark:* Conjecture C just states that the absorptive part
slightly above the threshold of **G** is precisely the absorption integral.
But here the absorption integral is defined even far from the threshold, and it is not certain that this integral will everywhere admit
an interpretation as an "absorptive part," i.e., as a difference
between two analytic continuations of the $S$-matrix. We shall indicate later what could prevent such an interpretation from holding.

Before proceeding further, we must show that Definition 6
makes sense. This is not quite obvious, since $\langle S_v \rangle_T$ is a distribution
and we have first to make a product of such distributions and then
integrate it. But according to Conjecture A and B, $\langle S_v \rangle_T$ is a
boundary value of an analytic function, and so will be the integrand
in Definition 6. Integrating the boundary value of an analytic function on a real cycle amounts to integrating this analytic function
on a suitable "complex diversion" of the above real cycle. Therefore, the absorption integral will be defined at all points where
such a diversion is possible. According to the Appendix, the bad
points of $M$ are the critical values of the mapping $\rho$ and of its
restrictions to the singular loci of the integrand; at all the other
points the diversion will be possible, so that the integral will be

defined, and will even be an analytic function of $p$. We thus are left with the purely geometrical problem of finding these critical values.

But first we should make more precise the singular structure of the integrand. Given a vertex $v$, any Landau singularity of $\langle S_v \rangle_T$ can be considered as the leading singularity of some multiple scattering process $G_v$. Let $G_*$ be the graph deduced from $G$ by replacing the elementary process occurring at $v$ by the multiple scattering process $G_v$. This graph $G_*$ will be called an "extension over $G$, with kernel $G_v$." Clearly $G$ is a contraction of $G_*$ (deduced from $G_*$ by contracting $G_v$), and one can define, in a way quite similar to Section 2, a canonical mapping

$$\rho_v : \mathbf{M}_* \to \mathbf{M}$$

from the space of $G_*$ to the space of $G$. All the considerations of Section 2 apply equally well to this mapping, with the role of the "internal lines" now played by the lines of $G_v$. As in Section 2, the critical points of $\rho_v$ are given by the Landau equations, which are precisely the equations giving the singularities of $\langle S_v \rangle_T$. In this way the analytic properties of the integrand $S_G = \prod_v \langle S_v \rangle_T$ can be formulated in complete analogy with Conjectures A, B, and C:

*Property A:* The distribution $S_G$ is equal to an analytic function everywhere on $\mathbf{M}$, except at the relevant Landau points of all extensions over $G$.

*Property B:* Near a smooth, relevant Landau point, the distribution $S_G$ is the boundary value of an analytic function. The "diversion" along which the boundary value is to be taken corresponds to an imaginary part "pointing upward from the threshold."

*Property C:* Near a smooth Landau point, leading Landau point of one extension $G_*$ and of no other, the discontinuity of the integrand is given by

$$S_G(\mathbf{p}) - S_G^{L_v}(\mathbf{p}) = \int_{\rho^{-1}_v(\mathbf{p})} S_{G^*}.$$

The main fact implied by Property A is the following: The singularities of the integrand are the apparent contours of mappings $\rho_v : \mathbf{M}_* \to \mathbf{M}$. Now let $L_v \subset \mathbf{M}$ be such a singularity. If it is a leading Landau singularity of $\rho_v$, then it is a manifold of codimension one (by Proposition 4 applied to $\rho_v$ instead of $\rho$), and the tangent space to that manifold is just the image of the tangent space to $\mathbf{M}_*$.

Therefore, projecting all this on $M$, i.e., considering the commutative diagram

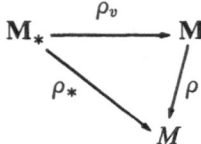

we see that the image (by $\rho$) of the tangent space to $L_v$ coincides with the image (by $\rho_*$) of the tangent space to $\mathbf{M}_*$, in particular, the apparent contour of $L_v$ will be part† of the apparent contour of $\mathbf{M}_*$.

This completes our task of finding the bad points of $M$; they are, besides the Landau points of $\mathbf{G}$, the Landau points of all graphs $\mathbf{G}_*$ which are extensions over $\mathbf{G}$. We thus have proved‡ the first part of

*Theorem A. The absorption integral $A_\mathbf{G}(p)$ is well defined, and equal to an analytic function, at all points of $M$ above the threshold of $\mathbf{G}$ except the Landau points§ of all graphs $\mathbf{G}_*$ which are extensions over $\mathbf{G}$ (including the trivial extension, i.e., $\mathbf{G}$ itself). Further, among these Landau points, one must retain only those where the $\alpha$ parameters of the Landau equations are non-negative for every line of the kernel of the extension.*

This last part of Theorem A follows from the fact that only the apparent contours of the relevant parts of $L_v$ matter. These relevant parts are defined by the non-negativity of the $\alpha$ parameters entering in the Landau equations of $\rho_v$. But the latter are just the $\alpha$ parameters associated to the lines of $\mathbf{G}_v$ in the Landau equations of $\rho_*$, as will be seen by the reasoning sketched below.

A local equation $l_v$ of $L_v$, the apparent contour of $\rho_v$, has its differential at any given Landau point $\mathbf{p}^c$ equal to a linear com-

---

†That it is only part of it is shown in the more detailed discussion of Section 6. But the extra points belong to the apparent contour of $\mathbf{M}$ so that they are singularities of the integral anyway.

‡There are some holes in the "proof" given here. For instance we have considered one Landau manifold $L_v$ separately, forgetting that several Landau manifolds of the integrand could conspire together to give singularities. A more complete treatment can be found in Ref. 1.

§Let us recall once more that our definition of Landau points is not quite the usual one (see the remark at the end of Section 2).

bination of differentials of mass shell equations:

$$dl_v = \sum_{i_v \in I_v} \alpha_{i_v} p^c_{i_v} \cdot dp_{i_v} \tag{1}$$

($I_v$ denotes the set of lines of $G_v$). The Landau equations just express that the above linear combination depends indeed only on $dp(p \in M)$. Similarly, a local equation $l_*$ of $L_*$, the apparent contour of $\rho|L_v$, has its differential at any given Landau point $p^c$ equal to a linear combination of differentials of $l_v$ and of mass shell equations

$$dl_* = dl_v + \sum_{i \in I} \alpha_i p^c_i \cdot dp_i \tag{2}$$

where I denotes the set of internal lines of **G**. Substituting equation (1) in (2) gives

$$dl_* = \sum_{i_* \in I_*} \alpha_{i_*} p^c_{i_*} \cdot dp_{i_*} \tag{3}$$

($I_* = I \cup I_v$ is the set of internal lines of $G_*$). Writing that (3) depends indeed only on $dp(p \in M)$ gives the Landau equations of $G_*$, with $\alpha$ parameters which are the same as in (1) as far as the lines of $I_v$ are concerned.

Theorem A did not exploit all the information contained in the Appendix which tells us not only where the integral is singular but also how to bypass its singularities by complex detours. To see which of the two possible detours works, we shall follow the convention, explained in the Appendix, of representing a detour by a vector transverse to the singularity. Here our integration manifold **M** has been diverted away from $L_v$ by a vector field transverse to $L_v$ pointing upward the threshold (Property B). According to Section 2, a local equation $l_v$ of $L_v$ will have the correct sign, i.e., will be positive above the threshold, if its differential is given by an expression (1) with non-negative $\alpha$ parameters. With such an $l_v$, the diversion of **M** will be described by a vector field with positive $dl_v$. Therefore, using equation (2), and remembering that $p^c_i \cdot dp_i$ (the differential of a mass shell equation) vanishes on **M**, we see that the projection of this vector field on $M$ will have a positive $dl_*$. The Appendix thus leads us to the following:

*Theorem B:* Let $L_* \subset M$ be the Landau manifold of a graph $G_*$ (extension over **G**) in the neighborhood of a Landau point which (a) is smooth (Definition 5) (b) is the projection of only one critical point, and (c) satisfies the relevance condition of Theorem A. *Then the absorption integrals $A_G$ defined on either side of $L_*$ are analytic* *continuations of each other along a complex detour with* $\operatorname{Im} l_*^c > 0$, *where* $l_*$ *is a local equation of* $L_*$ *with a sign chosen such that* $dl_* = \sum_{i_* \in l_*} \alpha_{i_*} p_{i_*}^c \cdot dp_{i_*}$ *with non-negative* $\alpha$'s *for every line of the kernel of the extension. This detour will be called the "natural detour for the absorption integral* $A_G$."

More generally, a "natural path" will be a succession of natural detours and real paths not meeting the singularities. It should be noticed that any path which is natural for the absorption integral is also natural for the scattering amplitude. Indeed, among the Landau points which are relevant for $A_G$ (those with $\alpha_{i_*} \geqslant 0$) will figure all the relevant Landau points of the scattering amplitude ($\alpha_{i_*} \geqslant 0$), and the "natural" detour given by Conjecture B coincides in that case with the natural detour of Theorem B. Consequently, the identification between absorption integral and absorptive part:

$$A_G = \langle S \rangle_T - \langle S \rangle_T^{(L)}$$

which was valid, according to Conjecture C, slightly above the threshold of **G**, can be continued analytically along any natural path of $A_G$, defining the analytic continuation of $\langle S \rangle_T^{(L)}$ along such a path.

A word of caution: The Landau loci of the graph **G** itself (even the "irrelevant" parts) will in general be impassable for the absorption integral $A_G$. To be more precise, any detour across such a singularity will give an analytically continued function which will no longer coincide with the absorption integral (it will in general be an integral over a *complex* cycle). See the remark AIII 3.2 of Ref. 1. The existence of such impassable barriers might make one question the universal validity of the interpretation of absorption integrals as absorptive parts.[†]

We shall complete Theorems A and B by a theorem giving the discontinuity of the absorption integral across a singularity. Let $L_*$ be a leading Landau locus of the graph $G_*$ (extension of **G**), and let $A_G(p)$ be the absorption integral at a point $p$ slightly above the threshold $L_*$. As we have seen, it can be deduced from the absorption integral under the threshold by analytic continuation along a natural detour. Let $A_G^{L_*}(p)$ now be the function obtained by analytic continuation along the opposite detour. We shall prove the following:

---

[†]One-loop graphs—the only ones studied in great detail up to now—are too simple to display such oddities, but the study of two-loop graphs might reveal interesting features.

*Theorem C: At a point p slightly above the threshold $L_*$, the discontinuity $A_G(p) - A_G^{L_*}(p)$ is equal to the absorption integral $A_{G^*}(p)$.*

Let us first notice that $L_*$, a leading Landau locus of $G_*$, can be considered as the apparent contour of $L_v$, a leading Landau locus of the extension. Therefore, both mappings $\rho_*$ and $\rho_v$ in the already drawn commutative diagram have the $S_+^1$ type, a fact which requires $\rho|L_v$ also to have the $S_+^1$ type (see Appendix II of Ref. 1). Thus, we are exactly in the conditions where the discontinuity formula of the Appendix can be applied. It gives

$$A_G(p) - A_G^{L_*}(p) = \int_{e_{(L_v)p}} [S_G(\mathbf{p}) - S_G^{L_*}(\mathbf{p})]$$

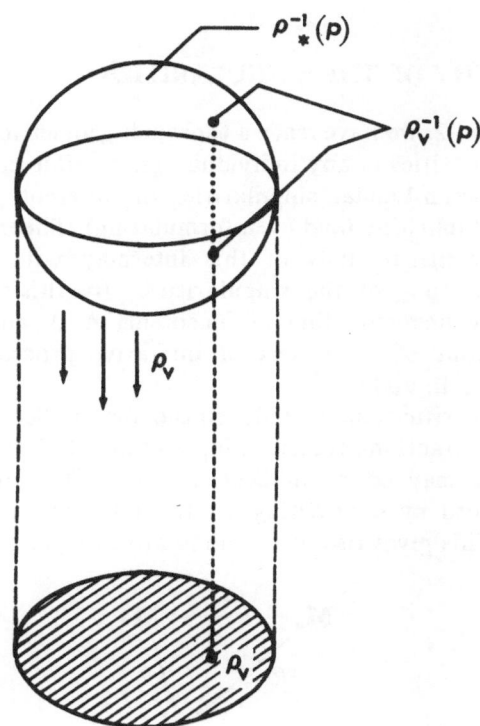

Fig. 11. The sphere $\rho_*^{-1}(p)$ can be considered as the cell $e_{(L_v)p} = \rho^{-1}(p) \cap \operatorname{Im} \rho_v$ "fibered" by the "vanishing sphere" $\rho_v^{-1}(p)$ (a zero-sphere on the picture, i.e., a couple of points, which "vanishes" when the two points come to coincide, on the equator of the sphere).

where the integration cell $e_{(L_v)_p}$ is the part of the fiber $\rho^{-1}(p)$ enclosed by the Landau locus $L_v$; since $L_v$ is the boundary of the projection $\rho_v$, this cell can be described also as the restriction to the fiber $\rho^{-1}(p)$ of the image of $\rho_v$; but Property C expresses the integrand as an integral of $S_{G_*}$ over the fiber of $\rho_v$: substituting this expression, we find

$$A_G(p) - A_G^{L_*}(p) = \int_{\rho_*^{-1}(p)} S_G.$$

which is exactly Theorem C. The above reasoning might be found easier to follow on Fig. 11, which shows all the cells and spheres involved.

## 6. HIERARCHY OF THE SINGULARITIES

In Section 2, we have made a thorough geometrical study of the Landau singularities of any individual graph. But in Section 5, the interplay between Landau singularities of different graphs was essential for establishing (and even formulating) Theorems A, B, and C. The geometrical study of this interplay will be called the "hierarchical study of the singularities." By this study, we shall gain a better understanding of Theorems A, B, and C, and also perhaps, a hint of some sort of interative procedure justifying Conjectures A, B, and C.

Take the situation, already considered in Section 5, of two successive contractions transforming a graph $G_*$ into $G$ and then $G$ (this last one may be, as in Section 5, the "elementary scattering graph" deduced by contracting all the internal lines, but this is irrelevant). This gives rise to a commutative diagram of mappings

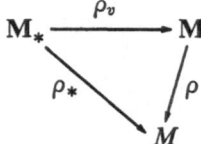

The main hierarchical problem consists in comparing the apparent contours of such mappings. We can forget about graphs and consider, quite generally, any commutative diagram of mappings of

manifolds

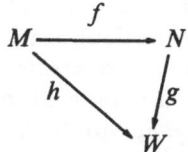

Let us ask about the critical character of $h$ at a point $x \in M$. Calling $y$ and $z$ the images of $x$ in $N$ and $W$, we easily verify that (a) if $x$ is not critical for $f$, it will be critical for $h$ if and only if $y$ is critical for $g$, and (b) if $x$ is critical for $f$ with type $S^1$, it will be critical for $h$ if and only if $y$ is critical for the mapping $g$ restricted to the apparent contour of $f$. Indeed, saying that $x$ is critical for $h$ means that the tangent space $T_x M$ does not project onto $T_z W$, and of course this is equivalent to saying that the image of $T_x M$ in $T_y N$ does not project onto $T_z W$. But this image is just (a) $T_y N$ itself if $f$ is regular, and (b) the tangent space to the apparent contour of $f$ if $f$ is singular of type $S^1$.

In the space $W$, the above result means that the apparent contour of $h$ is made up of (a) part of the apparent contour of $g$ (the part coming from the critical points of $g$ which belong to the image of $M$), and (b) the apparent contour of the apparent contour of $f$.

We try to illustrate this in Fig. 12, where $W$ is the plane of the picture, $N$ is a sphere, and $g$ is the projection on the plane of the picture. $M$ is not represented, but we suppose that its projection on $N$ is the upper hemisphere, and its apparent contour the equator of $N$. This projection is represented by the shady zone on the picture.

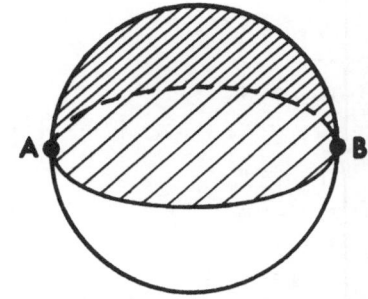

Fig. 12. An *effective contact* occurs at points A and B.

The apparent contour of $M$ in $W$ consists of two parts: (1) the upper half circle, half the apparent contour of the sphere; (2) the ellipse, apparent contour (here projection) of the equator.

Figure 12 also shows an interesting feature: the two branches of the apparent contour are tangent to each other. This is not astonishing since the tangents to both branches at an intersection point such as $A$ are the projections of the same space, namely the tangent space to $M$. This quite general situation will be called "effective intersection" or "effective contact" of two Landau curves. To be more precise, let us give the following:

*Definition 7.* We say that the apparent contours of $g$ and $h$ have an *effective intersection* at $z \in W$ if $z$ is the image of some point $y \in N$ which belongs at the same time to the apparent contour of $f$ and to the critical set of $g$.

There can exist many types of effective intersections. The most interesting is the "effective contact," defined by the following Proposition (proved in Appendix II of Ref. 1):

*Proposition 7.* Let $f$ and $g$ have singularities of the type $S^1_+$, and let $y$ be a point of $N$ where the apparent contour of $f$ and the critical set of $g$ intersect transversally. Then, if $x \in M$ is the critical point of $f$ which gives the critical value $y$, $x$ will also be critical for $h$, with a corank which may take the value 1 or 2. If this corank is 1, local coordinates

$$x_1, x_2, \ldots, x_m \qquad \text{in } M$$
$$y_1, y_2, \ldots, y_n \qquad \text{in } N$$
$$z_1, z_2, \ldots, z_w \qquad \text{in } W$$

can be chosen, such that the mappings $f, g, h$ read

$$f: \begin{cases} y_1 & = x_1 \\ y_2 & = x_2 \\ \cdots\cdots \\ y_{n-1} = x_{n-1} \\ y_n & = x_n^2 + x_{n+1}^2 + \cdots + x_m^2 \end{cases}$$

$$g: \begin{cases} z_1 & = y_1 \\ z_2 & = y_2 \\ \cdots\cdots \\ z_{w-2} = y_{w-2} \\ z_{w-1} = y_{n-1} + y_n \\ z_w & = y_{w-1}^2 + y_w^2 + \cdots + y_{n-1}^2 \end{cases}$$

$$h = g \circ f \begin{cases} z_1 & = x_1 \\ z_2 & = x_2 \\ \cdots\cdots\cdots \\ z_{w-2} = x_{w-2} \\ z_{w-1} = x_{n-1} + x_n^2 + x_{n+1}^2 + \cdots + x_m^2 \\ z_w & = x_{w-1}^2 + x_w^2 + \cdots + x_{n-1}^2 \end{cases}$$

*Definition 8*: *An effective intersection is called an effective contact if it corresponds to the situation described in Proposition 7.*

On the local model given by Proposition 7, it is easy to calculate the apparent contour of $h$: it consists of two branches tangent to each other, represented on Fig. 13. The shady zone on Fig. 13 represents the image of $h$. Locally, the situation is exactly the same as in the neighborhood of the point $A$ (or $B$) of Fig. 12.

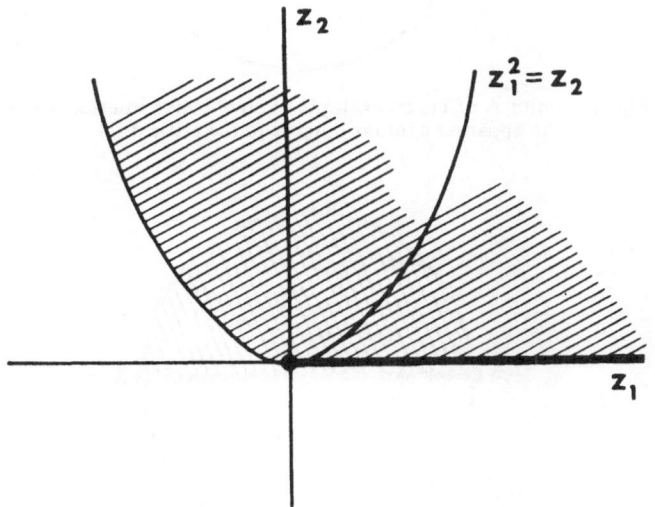

Fig. 13. Local model of an effective contact.

Let us comment upon the two conditions required by Proposition 7, namely, (a) the apparent contour of $f$ and the critical set of $g$ intersect transversally, and (b) the corank of $h$ is 1. Figures 14 and 15 illustrate how these conditions could be violated. In Fig. 14, the upper hemisphere of Fig. 12 has been replaced by a smaller zone whose boundary is tangent to the great meridian circle, critical set of the sphere, thus violating Condition 1. Figure 15 is just Fig. 12

seen from a different angle, such that the equatorial plane is perpendicular to the plane of the picture. Then the tangent to the equator at $A$ is no longer projected on a line but on a point, and this means that the corank of $h$ is 2. Notice that these are unstable situations which one can get rid of by arbitrarily small deformations.

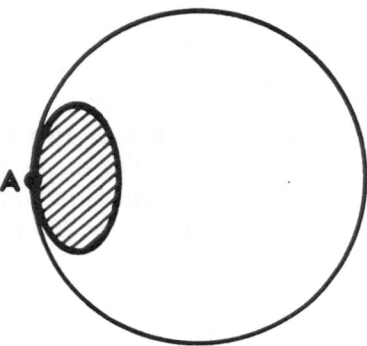

Fig. 14. Point A offers an example of an effective intersection with the apparent contour tangent to the critical set.

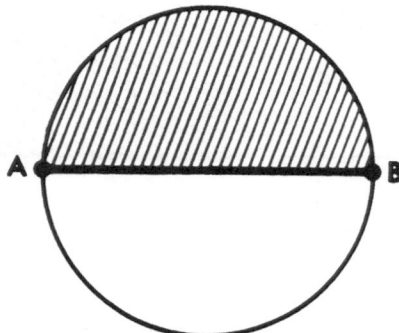

Fig. 15. Points A and B offer examples of effective intersections with corank 2.

In our problem of graphs, we already have at our disposal a simple criterion (Proposition 4) ensuring that $f$ and $g$ are of the $S_+^1$ type, so that only the above two conditions need verification in order to ensure that the effective intersection is an effective contact. The corresponding equations can be written (see I.3.1 of Ref. 1) and are easily exploited in each special case, but I have found no criterion as simple and general as that of Proposition 4 to ensure their fulfilment.

A curious application of the "effective contact" notion is made in Ref. 1 (see II 3.3). It starts from the remark that among the two branches of parabola shown on Fig. 13, only one is relevant (Definition 2), namely, the one on the left of the picture. The one inside the shady zone cannot be relevant since it violates Proposition 5 (Section 2). Now consider a multivalued analytic function of two complex variables $z_1, z_2$ which is singular on the straight line $z_2 = 0$ and on the parabola $z_1^2 = z_2$. Its different "determinations" are deduced from one another by analytically continuing along the different "loops" encircling the singularities. Now suppose that this function is not singular on the irrelevant branch of the parabola. In particular, a loop around that branch will not change its determination. But simple topological considerations show† that this loop can be deformed into some combination of loops around the straight line and the relevant branch of the parabola. Looking at this combination, one sees that the triviality of the action of the original loop is equivalent to the following condition

$$\text{Disc}_{L_*} (\text{Disc}_L f) = \text{Disc}_{L_*} f$$

where $f$ is the original analytic function, $\text{Disc}_L$ is the discontinuity around the left-hand branch of the straight line, and $\text{Disc}_{L_*}$ is the discontinuity around the left-hand branch of the parabola. Now if $f$ is interpreted as the $S$-matrix element and its discontinuities as absorption integrals, then the content of the above relation is exactly that of Theorem C. To summarize, purely topological considerations around an effective contact have shown us the equivalence of the following two properties:

(1) the $S$-matrix is a single-valued analytic function around the irrelevant parts of $L_*$;

(2) its absorptive parts around the relevant parts are related by

$$A^L - (A^L)^{L_*} = A^{L_*}$$

The first is part of Conjecture A for the graph $G_*$. The second leads to Conjecture C for $G_*$ if Conjecture C holds true for $G$ and $G_v$ (just follow the proof of Theorem C).

---

†For the learned readers, it means calculating the "fundamental group" of the complement of the line and the parabola.

Let us conclude this section with another remark. We have announced at the end of Section 2—and it must be clear now—that the "Landau singularities of a graph" are not just its "leading" Landau singularities (in the usual sense of this word) but also include some of the leading singularities of its contracted graphs. For instance, in the case of an effective contact, the Landau locus of $G_*$ is not just the parabola of Fig. 13, but also the right-hand part of the straight line, leading Landau locus of the contracted graph $G$. Failure to recognize this has led in the past to an incorrect formulation of Theorem A, to which its authors have recently given a counter example (Ref. 9). The decisive feature of that counter example was precisely its "effective contact" behavior. This may sound strange, since the notion of effective contact was already known before.† But the only examples treated to that time concerned contraction of one line of the graph $G_*$ (i.e., $G_v$ was a one-particle exchange graph), a case which is exceptional in the sense that $\rho_v \colon M_* \to M$ is a mapping from a space to another of greater dimension, and the "apparent contour" just means the projection. Imagine Fig. 12 on which the projection of the initial space would no longer be the hemisphere but the equator, and you will understand what makes such an effective contact different. Analytically, just replace the model for $f$ in Proposition 7 by

$$f \colon \begin{cases} y_1 & = x_1 \\ y_2 & = x_2 \\ \cdots\cdots \\ y_{n-1} & = x_{n-1} \\ y_n & = 0 \end{cases} \qquad (n-1 = m)$$

and you will see that the apparent contour of $g \circ f$ is only the parabola $\{z_1^2 = z_2\}$.

## APPENDIX: ANALYTICITY OF "ALMOST REAL" INTEGRALS

We shall develop here the mathematical tools needed in Section 5 for studying the analytic properties of absorptive parts, as given by the "Cutkosky" integral representation. We begin with a very naive version of J. Leray's Residue Theory, a basic tool for studying in general the analytic properties of multi-dimensional integrals.

---

† "Effective intersection" was the expression used.

## A1. A glimpse into Leray's Residue Calculus

Our study will be centered around integrals of the type

$$I = \int f(x_1, x_2, \ldots, x_n) \delta[s(x_1, x_2, \ldots, x_n)] dx_1 \, dx_2, \ldots, dx_n$$

where $f$ is an analytic function of $n$ real arguments, $\delta$ is the Dirac "$\delta$-function," and $s$ is a real analytic function which never vanishes together with its gradient. Then the set of points where it vanishes is a real analytic $(n-1)$-dimensional submanifold $S \subset \mathbb{R}^n$ (Theorem 2 of Ref. 2). We shall suppose that this submanifold is compact, so that the integral will converge without any trouble. One can consider, in a complex neighborhood of $S$, the "complexification" $S^c$ of $S$ given by the equation $s^c(z_1, z_2, \ldots, z_n) = 0$, where $s^c$ is the complexification of $s$, i.e., the complex analytic function of complex variables whose Taylor development at a real point is just that of $s$. Of course, the gradient of $s^c$, like that of $s$, will not vanish on $S$, so that $S^c$ will, in a sufficiently small neighborhood of $S$, be a complex analytic $(n-1)$-dimensional† submanifold of $\mathbb{C}^n$, depending only upon $S$ and not on the equation $s$ chosen.‡

Leray's residue theorem, which we shall prove below, tells that the above integral can also be written as

$$I = \frac{1}{2\pi i} \int_{\delta S} \frac{f^c(z_1, z_2, \ldots, z_n)}{s^c(z_1, z_2, \ldots, z_n)} \, dz_1 \wedge dz_2 \wedge dz_3 \wedge \cdots \wedge dz_n$$

where $f^c$ is the complexification of $f$ and $\delta S$ is an $n$-dimensional cycle§ encircling $S$ in $\mathbb{C}^n - S^c$, called the *coboundary of S*.

*Example:* $n = 1$, $s(x) \equiv x$. Then $I = \int f(x)\delta(x)dx = f(0)$, $S = \{0\}$. The ordinary residue theorem tells that $I = (1/2\pi i) \int_{\delta\{0\}} f^c(z)dz/z$, where $\delta\{0\}$—the coboundary of the origin—is a small positive circuit encircling the origin.

*Construction of the Coboundary.* A neighborhood of $S$ can be foliated by an analytic family $(\lambda_x)_{x \in S}$ of line segments $\lambda_x$ (the leaves

---

†Complex $(n-1)$-dimensional = real $(2n-2)$-dimensional.

‡All this reasoning concerning the "complexification" of a submanifold could of course be made for a submanifold of arbitrary codimension—not only codimension 1 as here.

§In the present connection, the word "cycle" can be understood as simply meaning "Compact oriented submanifold."

of the foliation) intersecting $S$ transversally at $x$. For instance $\lambda_x$ can be defined as the line segment orthogonal to $S$ at $x$. The complexification of every leaf $\lambda_x^c$ can be identified canonically with a small domain of the complex plane $\mathbb{C}$, with $x$ identified with the origin. Indeed, the function $s^c: \lambda_x \to \mathbb{C}$ (equation of $S^c$), realizes this identification. Then one can encircle $x$ in $\lambda_x^c$ by a small positive circuit depending continuously on $x$, and when $x$ runs over $S$, this small positive circuit generates an $n$-dimensional cycle in $\mathbb{C}^n - S^c$. This is the coboundary $\delta S$. Thus the coboundary $\delta S$ is topologically the product of a small positive circuit around the origin in $\mathbb{C}$ by the cycle $S$, and can be given the product orientation.†

*Proof of the Residue Theorem.* From the implicit function theorem one easily deduces that near every point of $S$ local analytic coordinates $\xi_1, \xi_2, \ldots, \xi_n$ can be chosen such that $\xi_1 = s$, whereas, the leaves $\lambda_x$ are the straight lines defined by giving constant values to $\xi_2, \ldots, \xi_n$. To show this, consider the mapping $\mu: \mathbb{R}^n \to S$ defined in the neighborhood of $S$ by associating to every point the origin $x$ of the leaf $\lambda_x$ to which it belongs. Now $\mu$ is an analytic mapping, and a retraction, i.e., it coincides with the identity on $S$. Therefore, on $S$, and in a sufficiently small neighborhood of $S$, its Jacobian matrix has maximum rank, so that $\mu$ is a submersion (Theorem 1 of Ref. 2), i.e., one can choose local coordinates $\xi_1, \xi_2, \ldots, \xi_n$ such that $\mu(\xi_1, \xi_2, \ldots, \xi_n) = (0, \xi_2, \ldots, \xi_n)$. In these coordinates the leaves $\lambda_x = \mu^{-1}(x)$ have just the above indicated form. Furthermore, $\xi_1$ is a local equation of $S$ and can be replaced by $s$ since the Jacobian determinant of the mapping $(\xi_1, \xi_2, \ldots, \xi_n) \mapsto (s, \xi_2, \ldots, \xi_n)$ is $\partial s/\partial \xi^1 \neq 0$.

Having covered the integration cycle $S$ with neighborhoods endowed with such local coordinates, we can cut it into a finite number of pieces $S_i$ each of which is contained in such a neighborhood. This gives for the integral a decomposition $I = \sum_i I_i$, where each piece can be rewritten in the corresponding $\xi$ coordinates:

$$I_i = \int_{S_i} f(x_1, x_2, \ldots, x_n) \delta[s(x_1, x_2, \ldots, x_n)] dx_1 \, dx_2 \cdots dx_n$$

---

†$S$ is understood to be oriented by the $(ds > 0)$-normal, i.e., a differential form $\omega$ of maximum degree on $S$ is positive if and only if $ds \wedge \omega$ is positive in $\mathbb{R}^n$.

$$= \int_{S_t} f(\xi_1, \xi_2, \ldots, \xi_n) \delta(\xi_1) \left| \frac{D(x_1, x_2, \ldots, x_n)}{D(\xi_1, \xi_2, \ldots, \xi_n)} \right| d\xi_1 \, d\xi_2 \, \cdots \, d\xi_n$$

$$= \int_{S_t} f(0, \xi_2, \ldots, \xi_n) \left| \frac{D(x_1, x_2, \ldots, x_n)}{D(\xi_1, \xi_2, \ldots, \xi_n)} \right| d\xi_2 \, \cdots \, d\xi_n$$

$$= \int_{S_t} f(0, \xi_2, \ldots, \xi_n) \frac{D(x_1, x_2, \ldots, x_n)}{D(\xi_1, \xi_2, \ldots, \xi_n)} d\xi_2 \wedge \cdots \wedge d\xi_n$$

where the suppression of the modulus bars | | on the Jacobian determinant and the replacement of the positive measure $d\xi_2 \cdots d\xi_n$ by the differential form $d\xi_2 \wedge \cdots \wedge d\xi_n$, are justified because, with $S$ oriented as already indicated, the above differential form is "positive" whenever the above Jacobian determinant is positive.

But now the ordinary one-variable residue theorem allows us to write

$$f(0, \xi_2, \ldots, \xi_n) = \frac{1}{2\pi i} \int \frac{f^c(\zeta_1, \xi_2, \ldots, \xi_n)}{\zeta_1} d\zeta_1$$

with the integral taken along a small positive circuit around the origin in the complex plane of $\zeta_1 = s^c$. Substituting this expression, we find

$$I_t = \frac{1}{2\pi i} \int_{\partial S_t} \frac{f^c(\zeta_1, \zeta_2, \ldots, \zeta_n)}{\zeta_1} \frac{D(z_1, z_2, \ldots, z_n)}{D(\zeta_1, \zeta_2, \ldots, \zeta_n)} d\zeta_1 \wedge d\zeta_2 \wedge \cdots \wedge d\zeta_n$$

$$\int_{\partial S_t} \frac{f^c(z_1, z_2, \ldots, z_n)}{s^c} dz_1 \wedge dz_2 \wedge \cdots \wedge dz_n$$

and summing up the different pieces $I_t$ gives precisely Leray's residue theorem.

## A2. Analyticity of a Real Integral

As an application of Leray's residue theorem, suppose that the functions $f$ and $s$ of **A1** depend analytically upon a real parameter $t$, thus defining a function

$$I(t) = \int f(x_1, x_2, \ldots, x_n, t) \delta[s(x_1, x_2, \ldots, x_n, t)] dx_1 \, dx_2 \, \cdots \, dx_n$$

Leray's residue theorem allows us to write

$$I(t) = \frac{1}{2\pi i} \int_{\partial S_t} \frac{f^c(z_1, z_2, \ldots, z_n, t)}{s^c(z_1, z_2, \ldots, z_n, t)} dz_1 \wedge dz_2 \wedge \cdots \wedge dz_n$$

The interesting fact is that the integral, originally defined on a cycle $S_t$ which varied with $t$, is now defined on a cycle $\delta S_t$ which can be chosen independent of $t$ for a sufficiently small variation of $t$. Indeed, the same family of line segments, and the same circuits, used for the construction of the coboundary, can serve for slightly different positions of the submanifold $S_t$. But obviously the integral, over a fixed cycle, of a function depending analytically of $t$, also depends analytically of $t$ (just integrate term by term the Taylor development of the integrand). Thus, we arrive at the conclusion that the above integral $I(t)$ depends analytically on $t$. In the proof we have needed the fact that $S_t$ is a compact submanifold (so that the coboundary $\delta S_t$ can be constructed) varying continuously with $t$ (so that the coboundary can be chosen the same for slightly different values of $t$). Precise hypotheses ensuring the fulfilment of these conditions are given below:

*Proposition I: Let S be the submanifold of $\mathbb{R}^{n+1}$ given by the analytic equation $s(x_1, x_2, \ldots, x_n, t) = 0$ (with $ds \neq 0$ on S); denote by $\tau \colon S \to \mathbb{R}$ the projection of S on the last coordinate axis t, and by $S_t = \tau^{-1}(t)$ the fiber of this projection. Then, if the mapping $\tau$ is proper without critical point, any analytic function $f(x_1, x_2, \ldots, x_n, t)$ integrated on $S_t$ with the measure $\delta[s(x_1, x_2, \ldots, x_n, t)]dx_1 \cdots dx_n$ yields an analytic function of t.*

Indeed, the absence of critical points (which can be expressed analytically by the nonvanishing of the differential $d_x s$) ensures that $S_t$ is a submanifold. Further, together with the proper character of $\tau$, it ensures that $S_t$ varies continuously with $t$.† This proper character also ensures that $S_t$ is compact, so that all the required conditions are satisfied.

In effect the above proposition can easily be generalized to the case of a manifold with arbitrary codimension (not only codimension 1 as above), using an iterated residue formula which generalizes the formula given in A1. Thus, one proves the following:

*Proposition I: Let S be the submanifold of $\mathbb{R}^{n+1}$ given by k analytic equations $s_i(x_1, x_2, \ldots, x_n, t) = 0, i = 1, 2, \ldots, k$ (with $ds_1 \wedge ds_2 \wedge \cdots \wedge ds_k \neq 0$ on S); denote by $\tau \colon S \to \mathbb{R}$ the projection of S on the last coordinate axis t, and by $S_t = \tau^{-1}(t)$ the fiber of this projection. Then, if the mapping $\tau$ is proper without critical points, any analytic*

---

†It even ensures—it is a well known topological result—that the projection $\tau$ is a "locally trivial fibration."

*function $f(x_1, x_2, \ldots, x_n, t)$ integrated on $S_t$ with the measure $\mu_t =$* ( $\prod\limits_{i=1,2,\cdots,k} \delta[s_i(x_1, x_2, \ldots, x_n, t)])dx_1 dx_2 \cdots dx_n$ *yields an analytic function of $t$.*

All the above discussion required $f$ to be analytic, more precisely holomorphic in a neighborhood of $S$. But what will happen if $f$ has singularities which meet $S$? Then the integral over $S_t$ will make no sense, unless we can avoid the singularities by moving $S_t$ a little in its complex neighborhood $S_t^c$. Such a small complex distortion of a real cycle will be called a *diversion*. Now we must point out that Leray's residue theorem, given here for a real integration cycle, also can be formulated in the complex case† with the same consequence that the integral is again analytic if the (diverted) integration cycle moves continuously with $t$. With these ideas in mind, we shall now investigate the problem of constructing suitable diversions. To simplify the exposition, the cycle $S$ will be replaced by a Euclidean space, but any reader with some knowledge of topology can easily adapt the arguments to the case of an arbitrary manifold.

## A3. Complex Diversions of Real Manifolds

The problem is to construct complex diversions of the real space $\mathbb{R}^n$, away from some real analytic submanifold $\Sigma \subset \mathbb{R}^n$ (or rather from its complexification $\Sigma^c \subset \mathbb{C}^n$). By a complex diversion we mean an imbedding of $\mathbb{R}^n$ in $\mathbb{C}^n$ defined by giving a small imaginary part to each point of $\mathbb{R}^n$, i.e.,

$$x = (x_1, x_2, \ldots, x_n) \longmapsto x + iy = (x_1 + iy_1, x_2 + iy_2, \ldots, x_n + iy_n)$$

with the $y_i$ some suitably chosen functions of $x$. For visualization, it will be convenient to consider these functions $y_i(x)$ as the coordinates of a real vector field on $\mathbb{R}^n$. Of course, this interpretation is not covariant (under curvilinear changes of coordinates in $\mathbb{R}^n$, the $y_i$ do not transform as the coordinates of a tangent vector), but it becomes covariant if the $y_i$ are *infinitesimal*. More precisely, consider a one-parameter family of diversions $x \to x + iy^\tau(x)$ $(y^0(x) = 0, \forall x)$ and define the infinitesimal generators of the

---

†With the $\delta$ function replaced by another concept: the "division by a differential form."

diversions by

$$Y_i(x) = \frac{\partial}{\partial \tau} y^\tau(x)\Big|_{\tau=0}$$

These infinitesimal generators are the coordinates of a real vector field $Y(x)$ with the correct covariance properties.

The reason for this can be seen in the following properties of infinitesimal motions in $\mathbb{C}^n$: Consider the transformation law of a tangent vector in $\mathbb{C}^n$; it is given by an $2n \times 2n$ matrix, the Jacobian matrix of the coordinate transformation. Now let us direct your attention to real coordinate transformations, where the new coordinates are real analytic functions (i.e., their Taylor developments at a real point have real coefficients) of the old ones. Then the Jacobian matrix at a real point reads $\left(\frac{J|0}{0|J}\right)$, i.e., it does not mix the real and imaginary coordinates and transforms both in the same way. Therefore, at a real point one can covariantly identify the real and pure imaginary tangent spaces, by the law

$$(x_1, x_2, \ldots, x_n; 0, \ldots, 0) \longmapsto (0, \ldots, 0; x_1, x_2, \ldots, x_n)$$

In particular, consider the one-parameter motion $z^\tau = x + iy^\tau$ defined by the above diversions. Its generator is the pure imaginary tangent vector $(0, Y)$; identifying it with the real tangent vector $(Y, 0)$ yields the announced result.

To push further the correspondence between infinitesimal complex diversions and real vector fields, let us notice that any smooth vector field with compact support can be considered as the infinitesimal generator of a one-parameter family of diversions. This easily follows from the well-known integrability properties of vector fields. With all these facts in mind, we now can come to our problem of diverting $\mathbb{R}^n$, away from some analytic submanifold. We prove the following:

*Lemma: Let $\Sigma$ be a compact real analytic submanifold of $\mathbb{R}^n$, with codimension 1. Let $x \mapsto x + iy^\tau(x)$ be a one-parameter family of diversions, reducing to the identity outside a compact neighborhood of $\Sigma$, and whose infinitesimal generator $Y(x)$ is a vector field transverse to $\Sigma$. Then, for all sufficiently small $\tau \neq 0$, the diversion sends $\mathbb{R}^n$ outside the complexification $\Sigma^c$ of $\Sigma$.* Indeed, call

$$\varphi: \mathbb{R}^{n+1} \longrightarrow \mathbb{C}^n$$

$$(x, \tau) \longmapsto x + iy^\tau(x)$$

the mapping defined by the one-parameter family of diversions. For $\tau = 0$, $\varphi$ is an immersion transverse to the submanifold $\Sigma^c$. Indeed, the image of its tangent mapping at a point $(x, 0)$ is generated by the real tangent space to $\mathbb{C}^n$ at $x$ plus the pure imaginary tangent vector $(0, Y(x))$, whereas the tangent space to $\Sigma^c$ is generated by all vectors of the form $(X, 0)$ or $(0, X)$, $X \in T_x \Sigma$ (tangent space to $\Sigma$ at $x$), and it is, therefore, clear that these two tangent spaces together generate the whole tangent space to $\mathbb{C}^n$ if $Y(x)$ is transverse to $T_x \Sigma$. From this transversality property we conclude (by Theorem 4 of Ref. 2) that $\varphi^{-1}(\Sigma^c)$ is a submanifold of $\mathbb{R}^{n+1}$ with codimension 2 (since 2 is the real codimension of $\Sigma^c$ in $\mathbb{C}^n = \mathbb{R}^{2n}$), i.e., with dimension $n - 1$. More exactly it is possible to restrict the variation of $\tau$ to a small enough interval around zero for that to be true.† But this manifold $\varphi^{-1}(\Sigma^c)$ must contain $\Sigma \times (\tau = 0)$, which is a manifold with the same dimension $n - 1$. Since the latter is topologically closed, it must be a connected component of the former, and even coincide with it if the interval of variation of $\tau$ has been sufficiently restricted. Thus, the points of $\Sigma \times (\tau = 0)$ are the only ones which are sent to $\Sigma^c$ by $\varphi$, and the Lemma is proved.

So the problem of diverting $\mathbb{R}^n$ away from $\Sigma^c$ has been reduced to the very easy one of constructing a vector field transverse to $\Sigma$. Obviously, such a vector field can always be constructed locally, and by standard topological arguments it also can be constructed globally if $\Sigma$ is orientable. Further, there exist two‡ homotopically inequivalent ways of constructing it (corresponding to the two opposite ways of crossing the hypersurface $\Sigma$), which will give rise to two classes of diversion. For instance, in the case $n = 1$, $\Sigma$ is a point on the real line $\mathbb{R}$, and the two classes of diversion correspond to a detour in the upper half plane or in the lower half plane.

The interest of the above considerations really starts with the following problem: Let the hypersurface $\Sigma_t \subset \mathbb{R}^n$ vary with some real parameter $t$. Can one divert $\mathbb{R}^n$ from $\Sigma_t^c$ continuously with respect to the parameter $t$? This problem amounts to constructing on $\mathbb{R}^n$ a one-parameter family of vector fields $Y_t(x)$, transverse to $\Sigma_t$, with a continous dependence on $t$. Equivalently, introducing in the

---

†Here we use the fact that $\varphi$ coincides with the identity outside a compact set.

‡Or rather $2^c$, where $c$ is the number of connected components of $\Sigma$.

space $\mathbb{R}^{n+1} \ni (x, t)$ the hypersurface $\Sigma$, trajectory of the family $\Sigma_t$ [i.e., $(x, t) \in \Sigma \iff x \in \Sigma_t$], we have to construct in $\mathbb{R}^{n+1}$ a smooth vector field transverse to $\Sigma$ and with no horizontal component (no component along the $t$-axis). This is clearly possible as long as the tangent hyperplane to $\Sigma$ is not vertical (see Fig. 16), i.e., as long as the projection $\tau$ of the manifold $\Sigma$ on the $t$-axis has no critical point. Near a critical point, a vector field transverse to $\Sigma$ will necessarily have a horizontal component, which can be taken constant in the neighborhood of the critical point, and interpreted as a complex distortion of the $t$-parameter. All these local results can be made globally valid if the projection $\tau$ is proper, and lead to

*Proposition II.* Let $\Sigma$ be a real analytic submanifold of $\mathbb{R}^{n+1}$, orientable, with codimension 1; denote by $\tau \colon \Sigma \to \mathbb{R}$ the projection of $\Sigma$ on the last coordinate axis $t$, and by $\Sigma_t = \tau^{-1}(t)$ the fiber of this projection. *Then, if the mapping $\tau$ is proper without critical points, $\mathbb{R}^n$ can be diverted away from $\Sigma_t^c$ continuously with respect to the real parameter $t$. If $\tau$ has isolated critical points corresponding to different values of $t$, $\mathbb{R}^n$ can be diverted away from $\Sigma_{t^c}^c$ continuously with respect to the complex parameter $t^c$ when this parameter runs along some suitable complex diversion of the straightline $\mathbb{R}$, avoiding the critical values.*

The hypothesis that the critical points correspond to different

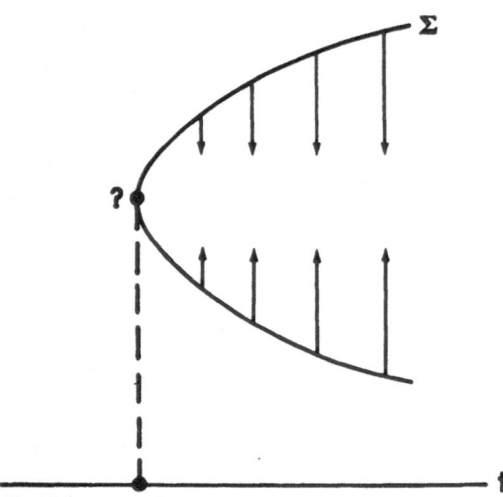

Fig. 16. Construction of a "vertical" vector field transverse to $\Sigma$.

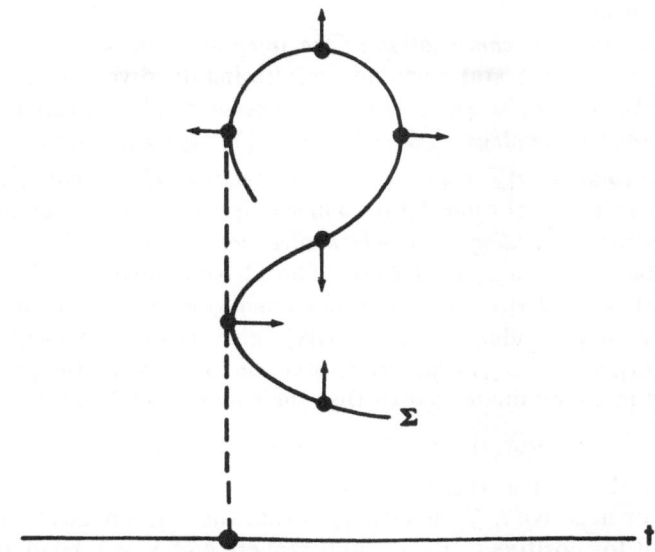

Fig. 17. Two critical points requiring contradictory diversions at the same critical value.

critical values was only meant to ensure that the globalization step does not require contradictory diversions of $t$, as the ones suggested by Fig. 17.

## A4. Singularities of an Almost Real Integral

We are now ready to complete the discussion at the end of Section A2, about diverting the integration cycle $S_t$ away from the singularities of the integrand. In Section A3, we have shown only how to divert a Euclidean space, but we have been careful to present things in a way which makes the generalization to manifolds obvious. Thus, Proposition II still holds if one replaces $\mathbb{R}^{n+1}$ by the submanifold $S$ of Proposition I: $S_t$ can be continuously diverted away from the submanifold $\Sigma_t^c \subset S_t^c$ on which the integrand is singular.[†] Taking the diverted $S_t$ as the integration cycle one obtains a well-defined integral[‡] which is an analytic function of $t$.

---

[†]This is supposing that the integrand is indeed singular on a submanifold; the case when it is singular on a union of submanifolds is treated in Ref. 1 (Appendix III).

[‡]If the integrand is multivalued, one has to choose its determination on the diverted $S_t$; this is of course always possible if $S_t$ is simply connected.

*To summarize:*

   The analytic continuation of the integral as an almost real integral (i.e., an integral over an infinitesimally diverted real cycle) is possible along the whole real axis, except at the critical values of the (proper) mappings $\tau: S \to \mathbb{R}$ and $\tau|\Sigma: \Sigma \to \mathbb{R}$. Moreover, if a critical value of $\tau|\Sigma$ comes from one isolated critical point, one can bypass this critical value by a complex detour in the lower or upper half plane, depending upon which diversion has been chosen for $S_t$.

   **An Important Special Case:** The $S_+^1$ singularity. Let $t^c$ be a critical value of $\tau|\Sigma$, coming from a nondegenerate quadratic critical point with null index ($S_+^1$ singularity). This means that local coordinates $(x_1, x_2, \ldots, x_p, t)$ can be chosen on $S$ ($\tau$ being the projection on the last coordinate $t$) such that the equation of $\Sigma$ reads

$$\sigma(x, t) \equiv x_1^2 + x_2^2 + \cdots + x_p^2 - t = 0$$

(taking the critical value $t^c$ to be 0).

   For negative $t$, $\Sigma_t$ is empty, so that there is no need to divert $S_t$. But for positive $t$, $\Sigma_t$ is a hypersphere and $S_t$ can be diverted in two ways, getting an "inward" or "outward" imaginary part. As clearly indicated by the vector fields of Fig. 18, the inward, respectively, outward diversion is obtained by a detour of $t$ in the upper, respectively, lower half plane. Notice that both diversions will correspond to the same determinations of the integrand *outside* the hypersphere. Thus, denoting by

   $e_{\Sigma_t}$ the ball† enclosed by the hypersphere $\Sigma_t$.

   $f^+(x, t)$, respectively, $f^-(x, t)$ the determination of the integrand, for $x$ inside the ball, deduced from its determination outside by the inward, respectively, outward diversion.

   $I^+(t)$, respectively, $I^-(t)$ the determination of the integral for $t > 0$, deduced from its negative $t$ determination by the detour in the upper, respectively, lower half plane; we obtain the following fundamental formula:

*Discontinuity Formula:*

$$I^+(t) - I^-(t) = \int_{e_{\Sigma_t}} [f^+(x, t) - f^-(x, t)]$$

where the integral is made, of course, with the same measure $\mu_t$ as in Proposition I.

---

   †This is called the "vanishing cell" in previous papers.

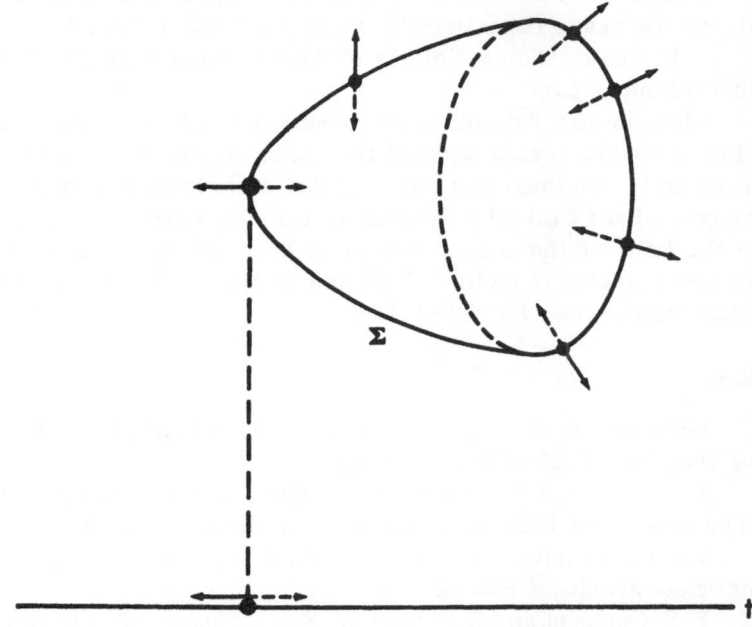

Fig. 18. Inward (- - - -→) and outward (←——) diversions near a
critical point of type $S^1_+$.

The above formula is valid whenever $f$ is summable near the
boundary $\sum_t$ of $e_{\sum_t}$. This is not the case if $f$ has a polar singularity
on $\sum_t$, but then $f$ is single-valued around $\sum_t$, and it is easily seen
that the difference between the inward and outward diversions is
just $\delta \sum_t$, Leray's coboundary of the sphere $\sum_t$. Therefore, applying
Leray's residue formula, we get in that case, supposing $f$ has a
simple pole $\hat{f}(x, t)/\sigma(x, t)$:

$$I^+(t) - I^-(t) = 2\pi i \int_{\sum_t} \hat{f}(x, t)\,\delta\,[\sigma(x, t)]$$

## REFERENCES AND COMMENTS

The substance of this article is taken from my thesis:
1. "Singularités des processus de diffusion multiple," *Ann. Inst.
Henri Poincaré*, Vol. VI, No. 2, p. 89–204, 1967. The presentation is
made more accessible, and improved in some respects: For instance,

in Section 2, the reasoning leading to Proposition 4 is made much simpler, thanks to the systematic study presented in Ref. 2.

2. F. Pham. "Some Notions of Local Differential Topology," (this volume p. 65).

Also, in the Appendix, the systematic use of vector fields allows a simpler presentation of the notion of diversion (détournement). Only two important points of Ref. 1 are left untouched here: the generalized Cutkosky–Steinmann relations, developed in Ref. 1 on the basis of the general notion of fiber-product, and the complications related to multiple lines and to the possibility of elastic scattering (Chapter III of Ref. 1).

## Section 1

For a review of the present situation about Landau singularities see the recent book of the following:

3. R.J. Eden, P.V. Landshoff, D.I. Olive, and J.C. Polkinghorne, "The Analytic S-Matrix," Cambridge University Press, (1966).

The recent interest in physical region singularities started with the short article of Ref. 4.

4. S. Coleman and R.E. Norton. *Nuovo Cimento* **38**: 438 (1965).

## Section 2

This section introduces my own way of looking at Landau singularities, to which I hope to convert people. In effect the idea that Landau singularities are apparent contours belongs to Professor R. Thom, but he instilled it into my mind in such a delicate way that I thought I was discovering it myself.

## Section 3

The conjectures in a vaguer form (A: analyticity except at the Landau singularities, B: prescriptions for bypassing the singularities, C: Cutkosky's rules) are ubiquitous in the papers dealing with the analyticity of the S-matrix. References can be found in Ref. 3.

## Section 4

The idea that the transition amplitudes should factorize for processes occurring in succession is the basis of Feynman's formulation of Quantum Mechanics (which gave birth to Feynman graphs, a fact that few people seem to remember!).

5. R.P. Feynman, *Rev. Mod. Phys.* **20**: 367 (1948).

An $S$-matrix version of this idea is given by

6. M.L. Goldberger and K.M. Watson, *Phys. Rev.* **127**: 2284 (1962).

A proof of the factorization for the double scattering process of Fig. 1 has been given, in the framework of axiomatic field theory, by

7. K. Hepp, *J. Math. Phys.* **6**: 1762 (1965).

## Section 5

This section can be considered as an elaboration of ideas contained, more or less implicitly, in the works of the so-called $S$-matrix axiomatists, mostly from Berkeley (H.P. Stapp) and Cambridge (J.C. Polkinghorne). That it should be possible to prove physical region analyticity with the help of these ideas is suggested by many examples worked out by the Cambridge group. The first was the following:

8. P.V. Landshoff and D.I. Olive, *J. Math. Phys.* **7**: 1464 (1966). (another exposition of this paper is given at the end of Ref 3).

## Section 6

The general notion of "effective contact" finds an illustration in Ref. 9.

9. P.V. Landshoff, D.I. Olive, and J.C. Polkinghorne, "The Hierarchical Principle in Perturbation Theory," Cambridge preprint, (1965).

## Appendix

The relevance of J. Leray's residue theory to the analytic study of integrals occurring in physics was pointed out some years ago by D. Fotiadi, M. Froissart, J. Lascoux, and myself. I do not give the references because the case of "almost real integrals" is much simpler: all the necessary material is given in this (almost) self-contained appendix. Here also many of the ideas are implicit in the works of the "$S$-matrix axiomatists."

# Some Notions of Local Differential Topology

F. PHAM

C.E.N.
Saclay, France

*and*

C.E.R.N.
Geneva, Switzerland

---

Exaggerating only slightly, one can say that local differential (or local analytic) topology is nothing but the art of making clever use of the implicit function theorem,† or equivalently, its simpler version called the inverse mapping theorem, which we shall recall to start with. But first let us agree on the following conventions, meant to simplify the exposition: We shall deal with mappings of a Euclidean space (called the source) into another Euclidean space (called the goal); it will not be necessary for these mappings to be defined in the whole of the source but only in some domain—which we shall not specify explicitly. The coordinates of a point in the source will be denoted by $x_i$, those of a point in the goal by $y_j$. A mapping $f$: $\mathbb{R}^m \to \mathbb{R}^n$ is therefore defined by

$$y_1 = f_1(x_1, x_2, \ldots, x_m); \quad y_2 = f_2(x_1, x_2, \ldots, x_m); \ldots$$
$$y_n = f_n(x_1, x_2, \ldots, x_m)$$

where $f_1, f_2, \ldots, f_n$ are $n$ real, numerical functions defined in some common domain of $\mathbb{R}^m$. If these functions are differentiable (respectively, analytic) in that domain, we shall say that the mapping is differentiable (respectively, analytic). It is understood that "differentiable" means $r$ times continuously differentiable, where $r$ is some integer $(1 \leq r \leq \infty)$ chosen once for all.

---

†Actually, there is another fundamental theorem in differential topology namely, *Sard's theorem* on the measure of the set of critical values of a mapping—but we shall not use it here.

## 1. FROM THE IMPLICIT FUNCTION THEOREM TO THE NOTION OF MANIFOLD

### 1.1. The Inverse Mapping Theorem

In this theorem the dimensions of the source and the goal are supposed to be the same.

*Let $f: \mathbb{R}^n \to \mathbb{R}^n$ be a differentiable* (respectively, analytic) *mapping. If at a point x* (in the domain of definition of *f*) *the Jacobian determinant* $\mathrm{Det}\|\partial f_j(x)/\partial x_i\|_{i,j}$ *does not vanish, then there exists a neighborhood $U_x$ of x mapped by f onto a neighborhood $V_y$ of y = f(x), such that* (a) $f|U_x: U_x \to V_y$ *is one-to-one, and* (b) *the inverse mapping* $(f|U_x)^{-1}$ (defined in $V_y$) *is differentiable* (respectively, analytic). We shall not prove this theorem, which is well known. One summarizes properties (a) and (b) by saying that *f* is a differentiable (respectively, analytic) *isomorphism*† between $U_x$ and $V_y$, or a local differentiable (respectively, local analytic) *isomorphism* of $\mathbb{R}^n$.

Since the following is a mere consequence of this theorem, we shall drop the word differentiable (respectively, analytic), simply speaking of mappings and isomorphisms.

### 1.2. Immersions and Submersions

In the general case where the source and the goal have unequal dimensions, one can also wonder what happens if the rectangular $m \times n$ matrix, defined by the partial derivatives $\partial f_j(x)/\partial x_i$, has a rank equal to its maximum possible value $\mathrm{Inf}\,\{m, n\}$.

*Case $m < n$:*

Write the matrix

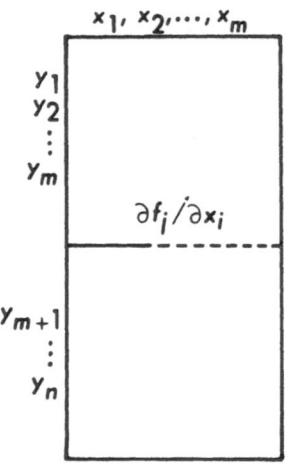

---

†One often says "diffeomorphism" instead of "differentiable isomorphism."

and suppose its rank is $m$, for instance its upper $m \times m$ minor has non-zero determinant. Then the mapping $(x_1, x_2, \ldots, x_m) \mapsto (y_1, y_2, \ldots, y_m)$ defined by the functions $f_1, f_2, \ldots, f_m$ is a local isomorphism of $\mathbb{R}^m$ since it satisfies the hypotheses of the inverse mapping theorem. Through this isomorphism of the source, the mapping $f$ is transformed into a mapping

$$g : \mathbb{R}^m \to \mathbb{R}^n ; (y_1, y_2, \ldots, y_m) \mapsto (y_1, y_2, \ldots, y_m, y_{m+1}, \ldots, y_n)$$

with $y_k =$ some function $g_k(y_1, y_2, \ldots, y_m)$, $k = m + 1, \ldots, n$. But now one notices that the mapping

$$(y_1, y_2, \ldots, y_m, y_{m+1}, \ldots, y_n) \mapsto (y_1, y_2, \ldots, y'_{m+1}, \ldots, y'_n)$$

defined by $y'_k = y_k - g_k(y_1, y_2, \ldots, y_m)$, $k = m + 1, \ldots, n$, is a local isomorphism of $\mathbb{R}^n$, since its Jacobian matrix

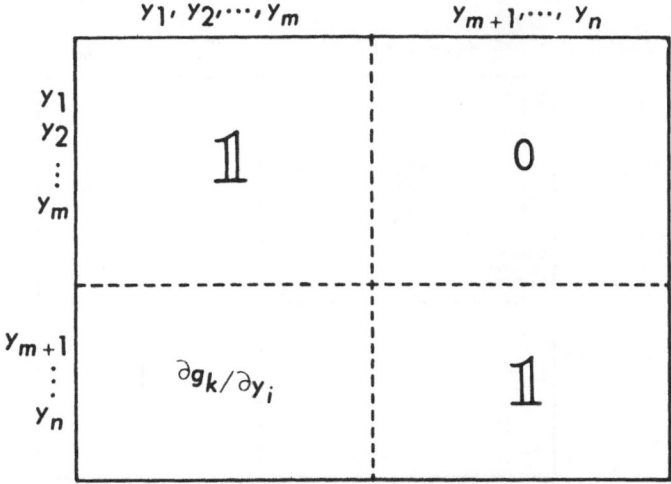

has its determinant equal to 1. Through this isomorphism of the goal, the above mapping $g$ is transformed into the mapping

$$\mathbb{R}^m \to \mathbb{R}^n$$

$$(y_1, y_2, \ldots, y_m) \mapsto (y_1, \ldots, y_m, 0, \ldots, 0)$$

which is just the canonical injection of the Euclidean space $\mathbb{R}^m$ into the bigger Euclidean space $\mathbb{R}^n$.

*To summarize: by suitable local coordinate transformations in the source and the goal, we have been able to transform locally the mapping* $f : \mathbb{R}^m \to \mathbb{R}^n$ *into a canonical inclusion of Euclidean spaces.*

*Case $m > n$:*
Write the matrix

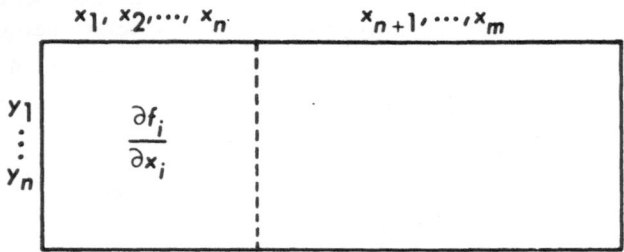

and suppose its rank is $n$, for example, its left $n \times n$ minor has non-zero determinant. Then the mapping $(x_1, x_2, \ldots, x_n, x_{n+1}, \ldots, x_m) \mapsto (y_1, \ldots, y_n, x_{n+1}, \ldots, x_m)$ defined by $y_i = f_i(x_1, x_2, \ldots, x_m)$ is a local isomorphism of $\mathbb{R}^m$ since its Jacobian matrix reads

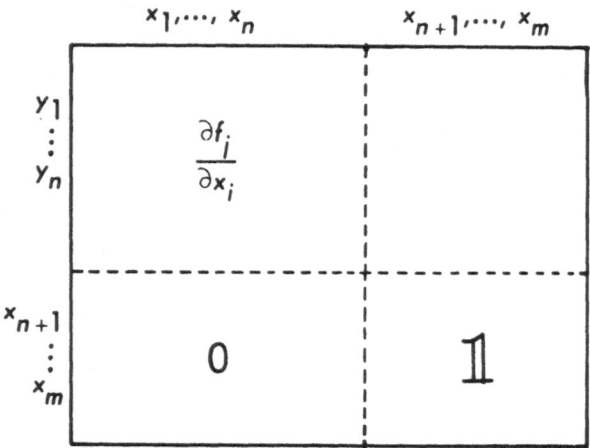

with a nonvanishing upper left $n \times n$ minor determinant. Through this isomorphism, the mapping $f$ is transformed into a mapping

$$\mathbb{R}^m \to \mathbb{R}^n$$

$$(y_1, y_2, \ldots, y_n, x_{n+1}, \ldots, x_m) \mapsto (y_1, y_2, \ldots, y_n)$$

which is just the canonical projection of the Euclidean space $\mathbb{R}^m$ onto the subspace $\mathbb{R}^n$

*To summarize: by a suitable local coordinate transformation in*

*the source, we have been able to transform locally the mapping $f$:
$\mathbb{R}^m \to \mathbb{R}^n$ into a canonical projection of Euclidean spaces.*

*Definition 1:*   Two mappings $f$, $g : \mathbb{R}^m \to \mathbb{R}^n$ are said to have
the same (local) type if there exist (local) isomorphisms $h$, $k$ of the
source and the goal such that the diagram

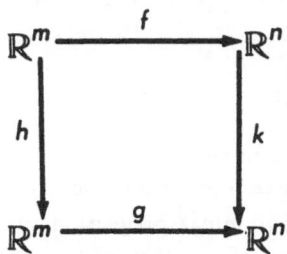

is commutative.

*Definition 2:*   A mapping is called an *immersion* (respectively,
a *submersion*) if it has the same local type as a canonical injection
(respectively, a canonical projection) of Euclidean spaces.

*Theorem 1:   A mapping $f : \mathbb{R}^m \to \mathbb{R}^n$ is an immersion $(m \leq n)$
or a submersion $(m \geq n)$ if and only if the rank of its Jacobian matrix
is equal to its maximum possible value* Inf$\{m, n\}$.† In fact, the "if"
part is just what has been proved above. The "only if" part is obvious, since canonical mappings of Euclidean spaces have this maximum rank property and the rank of a matrix cannot change when
it is multiplied (in the source and in the goal) by nonsingular square
matrices.

*Example 1a:*   Consider the parametric curve $\mathbb{R} \to \mathbb{R}^2$ de-
fined by $y_1 = x^2$, $y_2 = x^3$. The Jacobian matrix reads $\begin{array}{|c|} \hline 2x \\ \hline 3x^2 \\ \hline \end{array}$,
whose rank is zero at the origin. Therefore this mapping is not, near
the origin, an immersion of the straight line into the plane. Geome-
trically this is reflected in the singular shape of the curve at the
origin, where it has a "cusp."

*Example 1b:*   On the contrary, the mapping $\mathbb{R} \to \mathbb{R}^2$ defined

---

†When $m = n$, both notions of immersion and submersion coincide with
that of local isomorphism.  .

by $y_1 = x^2$, $y_2 = x^3 - x$ is everywhere an immersion since its

Jacobian matrix $\boxed{\begin{matrix} 2x \\ 3x^2 - 1 \end{matrix}}$ always has rank 1. Notice that the

corresponding curve has a double point (the point $y_1 = 1$, $y_2 = 0$ comes from the two values $x = \pm 1$), so that this mapping has not, *globally*, the same type as a canonical injection of the line into the plane.

## 1.3.  Manifolds

Let us come back to the case $m > n$, and consider a submersion $f: \mathbb{R}^m \to \mathbb{R}^n$. We have seen that locally this mapping could be transformed into a canonical projection of Euclidean spaces by a suitable coordinate transformation of the source alone. Now let us confine our attention to the set $f^{-1}(0)$ of those points in $\mathbb{R}^m$ which $f$ sends to the origin of $\mathbb{R}^n$. By the above coordinate transformation, $f^{-1}(0)$ is transformed into $\mathbb{R}^{m-n}$, the "orthogonal complement" of $\mathbb{R}^n$ in $\mathbb{R}^m$

To summarize: the subspace $f^{-1}(0) \subset \mathbb{R}^m$ can be locally transformed, by a suitable coordinate transformation of the ambient space $\mathbb{R}^m$, into a linear subspace $\mathbb{R}^{m-n} \subset \mathbb{R}^m$

*Definition 3:*   Two subsets $S, S' \subset \mathbb{R}^m$ are said to have the same (local) type if there exists an isomorphism (a local isomorphism) of $\mathbb{R}^m$ which transforms $S$ into $S'$.

*Definition 4:*   A subset $S \subset \mathbb{R}^m$ is called a submanifold of codimension $n$ (or of dimension $m - n$) if near each of its points it has the same local type as a linear subspace $\mathbb{R}^{m-n} \subset \mathbb{R}^m$.

*Theorem 2:   A subset $S \subset \mathbb{R}^m$ is a submanifold of codimension n if and only if it coincides in the neighborhood of each of its points with the set of solutions of n differentiable (respectively, analytic) equations $f_1(x) = 0$, $f_2(x) = 0, \ldots, f_n(x) = 0$ such that the matrix $\| \partial f_j / \partial x_i \|$ has rank n everywhere on S.*

In fact, the "if" part is just what has been shown at the beginning of this subsection. Notice that saying $S$ is the set of points satisfying $n$ equations $f_1, f_2, \ldots, f_n$ is the same thing as saying it is the reciprocal image of the origin through the mapping $f: \mathbb{R}^m \to \mathbb{R}^n$.

To prove the "only if" part, one just has to notice that $\mathbb{R}^{m-n}$ can be given in $\mathbb{R}^m$ by $n$ equations $x_1 = 0, x_2 = 0, \ldots, x_n = 0$ whose Jacobian matrix obviously has this maximum rank property, a property which cannot be changed by any isomorphism of $\mathbb{R}^m$.

Let us emphasize one important feature of submanifolds. Restricting to $S$ the above local isomorphism of the ambient space $\mathbb{R}^m$, we get a *homeomorphism* (i.e., a one-to-one continuous map, whose inverse is also continuous) between $S$ (in the neighborhood considered) and a Euclidean space $\mathbb{R}^{m-n}$. This leads us to define the abstract notion of "manifold," keeping the above feature of $S$ without making reference to any "ambient Euclidean space."

*Definition 5:* An *m-dimensional topological manifold* is a Hausdorff topological space $M$, every point of which has a neighborhood $U$ homeomorphic to a domain $E$ of $\mathbb{R}^m$. Such a homeomorphism $h: U \to E$ is called a *local chart*† of the manifold, and $U \subset M$ is the *domain* of the local chart $h$. A family of local charts whose union of domains is the whole of $M$ is called an *atlas* of $M$. An atlas is called *differentiable* (respectively, *analytic*) if any two of its local charts $h$, $h'$ are "differentiably (respectively, analytically) related," i.e., if the mapping $h' \circ h^{-1}: \mathbb{R}^m \to \mathbb{R}^m$ is differentiable (respectively, analytic) in its domain of definition.

*Definition 6:* A *differentiable* (respectively, analytic) *structure* on a topological manifold $M$ is defined by a differentiable (respectively, analytic) atlas of this manifold, with the convention that two such atlases define the same structure on $M$ if and only if the local charts of one are differentiably (respectively, analytically) related to the local charts of the other.

Notice that a "submanifold of a Euclidean space," in the sense of Definition 4, has indeed a differentiable (respectively, analytic) structure: if one restricts to $S$ the local differentiable (respectively, analytic) isomorphisms of the ambient Euclidean space involved in Definition 4, one gets a system of local charts of $S$ which are differentiably (respectively, analytically) related.

---

†The $m$ numbers $h_1(x), h_2(x), \ldots, h_m(x)$, coordinates of the point $h(x)$ in $\mathbb{R}^m$, are called *coordinates of the point* $x \in M$ *in the local chart $h$*.

*Example 2. Grassmann manifolds:* Let $G_m^n$ stand for the set of all $m$-dimensional vector subspaces of $\mathbb{R}^{m+n}$, i.e., all $m$-planes going through the origin in $(m + n)$-dimensional Euclidean space. We shall show how this set can be given the structure of an $(mn)$-dimensional analytic manifold, called the Grassmann manifold $G_m^n$.

First one must define a topology on $G_m^n$. This is easily done. For instance, $G_m^n$ can be made into a metric space by defining the "distance" between two $m$-planes as being their "angle"† in $\mathbb{R}^{m+n}$, and one can take the topology induced by this metric. Now let us show that any point in $G_m^n$ has a neighborhood homeomorphic to a domain of $\mathbb{R}^{mn}$.

Let $\prod^m$ be some point of $G_m^n$, i.e., some $m$-dimensional vector subspace of $\mathbb{R}^{m+n}$, and let $\prod^n$ be some $n$-dimensional vector subspace of $\mathbb{R}^{m+n}$, transverse to $\prod^m$.‡ Call $L_m^n(\prod^n)$ the subset of $G_m^n$ consisting of all those $m$-planes which are transverse to $\prod^n$. It is clear that $L_m^n(\prod^n)$ is a neighborhood of the point $\prod^m$ since any $m$-plane sufficiently close to $\prod^m$ will also be transverse to $\prod^n$. Now we can construct a homeomorphism between $L_m^n(\prod^n)$ and $\mathbb{R}^{mn}$, thanks to the following remark: Saying that an $m$-plane $\prod^{\prime m}$ is transverse to $\prod^n$ is equivalent to saying that $\prod^{\prime m}$ is isomorphic to $\prod^m$ through the linear projection "parallel to $\prod^n$." Therefore, $\prod^{\prime m}$ can be identified with the graph of a linear mapping $\prod^{\prime m} : \prod^m \to \prod^n$ Choosing a basis on $\prod^m$ and on $\prod^n$ one can represent any such linear mapping by an $m \times n$ matrix, i.e., by $mn$ numbers. This is the announced one-to-one correspondence between $L_m^n(\prod^n)$ and $\mathbb{R}^{mn}$. It depends only on $\prod^m$, $\prod^n$, and the basis chosen for them. It is easy to see that this correspondence, as well as its inverse, is continuous, so that it is indeed a homeomorphism. We leave it to the reader as an exercise to prove that the above defined local charts of $G_m^n$ are analytically related to each other, i.e., that the $mn$ numbers associated to an $m$-plane $\prod^{\prime m}$ by any particular choice (of $\prod^m$, $\prod^n$, and of bases on $\prod^m$, $\prod^n$) are analytic functions§ of

---

†The task of defining the "angle" between two $m$-planes in $(m + n)$-dimensional Euclidean space is left to the imagination of the reader.

‡Two vector subspaces $V$, $V'$ of a vector space $W$ are said to be *transverse to each other* if $V + V' = W$. Notice that in the special case where $\dim V + \dim V' = \dim W$ (as is the case here), this condition is equivalent to saying that $V$ intersects $V'$ only at the origin.

§In effect one even verifies thet they are rational functions, so that these charts define what is called an *algebraic structure* on $G_m^n$.

the $mn$ numbers associated to the same $\prod'^m$ by any other choice. This will end the proof that $G_m^n$ has the structure of an $(mn)$-dimensional analytic manifold.

Notice that the so-called *projective space* $P^n$ [the set of all directions in $(n + 1)$-dimensional Euclidean space] is just a special case of a Grasmann manifold:

$$P^n = G_1^n$$

*Definition 7:* Given two differentiable (respectively, analytic) manifolds $M$ and $N$, a *differentiable* (respectively, *analytic*) *mapping* of $M$ into $N$ is a continuous mapping $f$ of the topological space $M$ into the topological space $N$ such that for any couple of local charts $h: M \to \mathbb{R}^m$, $k: N \to \mathbb{R}^n$ the mapping $k \circ f \circ h^{-1}: \mathbb{R}^m \to \mathbb{R}^n$ is differentiable (respectively, analytic) in its domain of definition.

[As a special case, considering the real line $\mathbb{R}$ as a manifold, one gets the notion of a *differentiable* (respectively, an *analytic*) *function* $f: M \to \mathbb{R}$].

If $f$ is one-to-one and its inverse is also a differentiable (respectively, analytic) mapping, $f$ is called a *differentiable* (respectively, *analytic*) *isomorphism*.

In all the following the word "manifold" will be used for a topological manifold with a differentiable or analytic structure, and we shall drop the word "differentiable" (or "analytic") when we speak of "mappings," "isomorphisms," etc.

### 1.4. Tangent Space and Tangent Bundle to a Manifold

The reader may wonder why we have bothered about defining manifolds since all the reasonings we want to do are local so that we would lose nothing by staying in ordinary Euclidean space. But even locally there is a conceptual difference between a manifold and a Euclidean space, namely, that the Euclidean space has a linear structure which is lost when one considers it as a manifold, allowing "arbitrary" changes of coordinates. This freedom in changing coordinates makes it useful to build up some concepts which are invariant through such changes, and this is where the notion of manifold comes in—at this stage the reader must clearly make the distinction between a point of the manifold, supposedly defined "in an intrinsic way" (for instance a point $\prod'^m$ in $G_m^n$ of example 2) and the *coordinates of that point* which are just a

convenient way of representing it (for instance the matrix representing the plane $\prod'^m$ in example 2).

One important intrinsic concept is that of the "tangent space
to a manifold," which represents, so to speak, "all that the manifold remembers about the linear structure of Euclidean space." Let
us proceed to define it.

Given an $m$-dimensional manifold $M$, and a point $x \in M$, let us
consider couples $(h; X_1, X_2, \ldots, X_m)$ consisting of a local chart $h$ of
$M$ around $x$ and an $m$-uple of real numbers $X_1, X_2, \ldots, X_m$. On the
set of all such couples, let us define the following equivalence
relation:

$$(h; X_1, X_2, \ldots, X_m) \sim (h'; X'_1, X'_2, \ldots, X'_m)$$

if and only if

$$X'_j = \sum_i \frac{\partial (h' \circ h^{-1})_j}{\partial \xi_i} (\xi) \, X_i$$

where $\partial (h' \circ h^{-1})_j(\xi)/\partial \xi_i)$ is the Jacobian matrix of the mapping
$h' \circ h^{-1} \colon \mathbb{R}^m \to \mathbb{R}^m$, evaluated at the point $\xi = h(x)$.

*Definition 8:* The equivalence classes of the above equivalence
relation are called *tangent vectors* to $M$ at $x$. If $X$ is such a tangent vector—the equivalence class of a couple $(h; X_1, X_2, \ldots, X_m)$—
one says that $X_1, X_2, \ldots, X_m$ are the *coordinates of the tangent
vector $X$ in the local chart $h$.* Thus the above formula relates the
coordinates of $X$ in some local chart to its coordinates in any other
local chart, and may be called the "transformation law for tangent
vector coordinates."

The set of all tangent vectors to $M$ at $x$ can obviously be made
into a vector space, defining the addition by the addition of coordinates in any chosen local chart $h$. This is clearly independent of
the local chart chosen, since the transformation law of coordinates
is linear. One thus gets an $m$-dimensional vector space, called the
*tangent space* to $M$ at $x$, and denoted by $T_x M$.

*TANGENT BUNDLE.†*

It is often useful to consider the (disjoint) union of all tangent
spaces at all points. On this set $TM = \bigcup_{x \in M} T_x M$ a topology can be
constructed which satisfies the following two requirements:

---

†This part can be skipped in first reading.

1. Let $\pi: TM \to M$ be the mapping which to any tangent vector $X$ associates the point of $M$ where this tangent vector is defined. We require this mapping $\pi$ to be continuous. In other words, for every open set $U \subset M$ the reciprocal image $\pi^{-1}(U) = \bigcup_{x \in u} T_x M$ must be open in $TM$.

2. We notice that if $U$ is the domain of a local chart $h: U \to E \subset \mathbb{R}^m$ there exists a one-to-one correspondence

$$\bar{h}: \pi^{-1}(U) \to U \times \mathbb{R}^m$$

simply defined by associating to the tangent vector $X$ at $x$ ($x \in U$) the couple $(x; X_1, X_2, \ldots, X_m)$, where $X_1, X_2, \ldots, X_m$ are the coordinates of $X$ in the local chart $h$. We require $\bar{h}$ to be a homeomorphism.

It is easily seen that the above two requirements completely define the topology of $TM$. Moreover, we see that any point of $TM$ has a neighborhood $\pi^{-1}(U)$ homeomorphic to $U \times \mathbb{R}^m$, with $U$ itself homeomorphic to an open set of $\mathbb{R}^m$: this shows that $TM$ is a $2m$-dimensional topological manifold. The reader will easily verify that the above constructed local charts of this manifold are related to each other through differentiable (analytic) transformations (differentiable of class $C^{r-1}$ if $M$ was differentiable of class $C^r$), defining on $TM$ a differentiable (analytic) structure. But $TM$ is also endowed with another very interesting structure, called a "fiber bundle" structure.

Since we shall meet other examples of "fiber bundles" we now explain this concept in its full generality.†

*Definition 9a:* A space $B$ is called a fiber space with base $M$, fiber $F$, and projection $\pi$, if there exists a mapping $\pi: B \to M$ such that $\pi^{-1}(x)$ is homeomorphic to a given space $F$, for all $x \in M$; $\pi^{-1}(x)$ is called the fiber above $x$.

Here, $TM$ is clearly a fiber space with base $M$, fiber $\mathbb{R}^m$ and the projection $\pi: TM \to M$ already considered. The fiber above $x$ is $T_x M$.

*Definition 9b:* A fiber space $B \overset{\pi}{\to} M$ is called *locally trivial* if any $x \in M$ has a neighborhood $U$ whose reciprocal image admits a

---

†All "spaces," "mappings," "isomorphisms," etc., will be understood as manifolds, mappings of manifolds, isomorphisms of manifolds, etc., but actually this is irrelevant, and one could deal as well with topological spaces, continuous mappings, and homeomorphisms.

homeomorphism $h: \pi^{-1}(U) \to U \times F$ "respectful of the projection $\pi$," i.e., such that the diagram

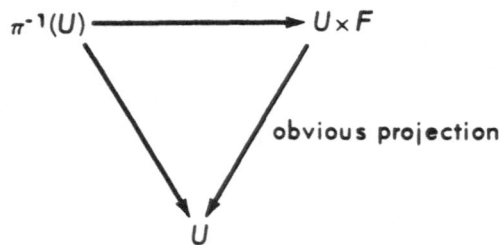

is commutative.

Such a homeomorphism $\bar{h}$ is called a "local trivialization of the fiber space."

Here the local trivializations of $TM$ are just the above constructed $\bar{h}$ associated to local charts $h$ of $M$.

*Definition 9c:* A locally trivial fiber space $B \xrightarrow{\pi} M$ is called a *fiber bundle* if it admits a privileged family of local trivializations "related to each other by transformations belonging to a given group $G$ of automorphisms of the fiber"—$G$ is called the *structural group of the bundle.* By "related to each other, etc." we mean the following: Let $\bar{h}$, $\bar{h}'$ be two local trivializations whose domains intersect, and restrict them to the intersection $U$ of these domains, then, since $\bar{h}$, $\bar{h}'$ are "respectful of the projection," $\bar{h}' \circ \bar{h}^{-1}: U \times F \to U \times F$ takes the form

$$(\bar{h}' \circ \bar{h}^{-1})(x, f) = (x, g_x(f))$$

where $g_x$ is some automorphism of $F$, depending on $x$. We require $g_x$ to belong to the given group of automorphisms $G$.

Here the structural group of $TM$ can be taken as the $m$-dimensional linear group. Indeed, if $\bar{h}(X) = (x; X_1, X_2, \ldots, X_m)$ and $\bar{h}'(X) = (x; X'_1, X'_2, \ldots, X'_m)$ are the trivializations associated to local charts $h$, $h'$ of $M$, then the vectors $(X_i)$ and $(X'_j)$ are related to each other by a linear transformation, namely the Jacobian matrix of $h' \circ h^{-1}$.

*Definition 9d:* A fiber bundle whose fiber is a vector space and structural group is the linear group is called a *vector bundle.* Thus the space $TM$ is a vector bundle over $M$. It will be called the *tangent bundle* of $M$.

*Definition 9e:* Given a fiber space $B \xrightarrow{\pi} M$, a mapping $\sigma : M \to B$ such that $\pi \circ \sigma = 1_M$ is called a *section* of the fiber space. The section is called continuous, differentiable, etc., if the mapping $\sigma$ is continuous, differentiable, etc. Here a section of $TM$ is a mapping $\sigma : M \to TM$ which to every point $x \in M$ associates an element $X$ of $T_x M$—it is called a *vector field* on $M$.

## 1.5.  Tangent Mapping: Transversality

Let $f : M \to N$ be a mapping of manifolds, and consider a local chart $h$ of $M$ around some point $x$, together with a local chart $k$ of $N$ around $y = f(x)$. Then $k \circ f \circ h^{-1}$ is a mapping of $\mathbb{R}^m$ into $\mathbb{R}^n$, and we can consider its Jacobian matrix

$$\left( \frac{\partial (k \circ f \circ h^{-1})_j (\xi)}{\partial \xi_i} \right)_{i,j} \qquad (\xi = h(x))$$

If $X_1, X_2, \ldots, X_m$ are the coordinates of a tangent vector $X \in T_x M$ in the local chart $h$, one easily verifies that the numbers

$$Y_j = \sum_i \frac{\partial (k \circ f \circ h^{-1})_j (\xi)}{\partial \xi_i} X_i$$

a) depend only on $X$ and not on the local chart $h$ chosen to define the coordinates $X_i$, and b) depend on the local chart $k$ exactly as the components of a tangent vector $Y \in T_y N$. Therefore the above Jacobian matrix defines a linear mapping $T_x f : T_x M \to T_y N$, called the *tangent mapping* to $f$ at $x$.

*Immersions and submersions. Transversality.* Since the notions of immersion and submersion introduced in Section 1. 2 for mapping of Euclidean spaces are purely local, they apply equally well to mappings of manifolds. Not a single word has to be changed to Definition 2. As regards Theorem 1, it can be reformulated as the following:

*Theorem 1: A mapping is an immersion* (respectively, a *submersion*) *if and only if its tangent mapping is injective* (respectively, *surjective*). Submanifolds of a given manifold $M$ are defined as in subsection 1. 3, Definition 4. Theorem 2 becomes the following:

*Theorem 2: A subset $S \subset M$ is a submanifold of codimension $k$ if and only if it coincides in the neighborhood of each of its points with the reciprocal image $f^{-1}(0)$ of the origin through a submersion $f : M \to \mathbb{R}^k$.*

Notice that if $S$ is a submanifold of $M$ there is a canonical immersion or "imbedding"† $i : S \to M$ defined by associating to each point $x \in S$ the same point considered in $M$. Obviously the tangent space $T_x S$ can be identified with a subspace of $T_x M$ (since the mapping $T_x i$ is injective). If $S$ is given by $f^{-1}(0)$ as in Theorem 2, this subspace $T_x S \subset T_x M$ is nothing but the *kernel* of $T_x f$, i.e., the set of tangent vectors which $T_x f$ sends to zero. This is obvious if one chooses in $M$ local coordinates in which $f$ reads as a canonical projection of Euclidean spaces. Notice that in arbitrary coordinates the kernel of $T_x f$ consists of those $(X_i)$ which satisfy the quite familiar equations

$$\sum_i \frac{\partial f_j}{\partial \xi_i} X_i = 0 \qquad j = 1, 2, \ldots, k$$

*Definition 10:* Given a submanifold $S \subset M$, and a point $x \in S$, the "transverse space of $S$ at $x$" is the quotient of vector spaces $T_x M / T_x S$. We shall denote it by $T_x(M/S)$.

Notice that if $f : M \to \mathbb{R}^k$ is a system of local equations of the submanifold $S$, $T_x f$ induces an isomorphism between the transverse spaces $T_x(M/S)$ and $T_0 \mathbb{R}^k$. Indeed, $T_x f : T_x M \to T_0 \mathbb{R}^k$ is surjective by hypothesis, and Ker $T_x f = T_x S$.

*Definition 11:* Given a submanifold $S \subset M$, a mapping $g : N \to M$ is said to be transverse to $S$ at the point $y \in N$ if either $g(y) \notin S$ or $g(y) = x \in S$ and the subspace of $T_x M$, image of the tangent mapping $T_y g$, is transverse to $T_x S$ (in the sense already mentioned in a footnote of Section 1.3, i.e., $T_x S$, together with that subspace, generate the whole of $T_x M$).

There is an obvious connection between this definition and Definition 10, namely,

*Theorem 3:* *The mapping g is transverse to S at y if and only if its tangent mapping sends $T_y N$ surjectively to the transverse space of S at $x = g(y)$, i.e., the composed mapping $T_y N \xrightarrow{T_y g} T_x M \to T_x(M/S)$ is surjective.* This obvious theorem immediately leads to the following:

---

†Generally speaking, an imbedding $j : N \to M$ is defined as a mapping which sends $N$ isomorphically onto a submanifold of $M$. Obviously any imbedding is an immersion. Conversely any immersion is locally an imbedding, but this need not be true globally (see Example 1b).

*Theorem 3:* Let $S$ be given near $x$ as $f^{-1}(0)$, the reciprocal image of zero through a submersion $f: M \to \mathbb{R}^k$. *Then the mapping* $g: N \to M$ *will be transverse to* $S$ *at* $y$ (supposing $g(y) = x \in S$) *if and only if the composed mapping* $f \circ g: N \to \mathbb{R}^k$ *is a submersion.*

Now noticing that $(f \circ g)^{-1}(0) = g^{-1}(S)$, and using Theorem 2, we immediately get

*Theorem 4:* *Near every point* $y$ *where the mapping* $g: N \to M$ *is transverse to* $S$, *the reciprocal image* $g^{-1}(S)$ *is a submanifold of codimension* $k$.

As an important special case, consider the case when $N$ is also a submanifold of $M$ and $g$ is the canonical imbedding. One then gets:

*Definition 11':* Two submanifolds $S, N \subset M$ *are said to be transverse* to each other at $x$ if either $x \notin S \cap N$, or $x \in S \cap N$ and the subspaces $T_x S$, $T_x N$ are transverse to each other in $T_x M$.

*Theorem 4':* *Near every point where the submanifolds* $S$, $N$ *are transverse to each other,* $S \cap N$ *is a submanifold of codimension* codim $S$ + codim $N$.

*Transverse Section of a Bundle.*† Given a bundle $B \overset{\pi}{\to} M$ with fiber $F$ and structural group $G$, an important concept is the "orbit" of a point $f \in F$, i.e., the set of points of the form $g(f)$, $g \in G$. Of course the special case of the tangent bundle is not interesting in this connection, since the only orbit is $F$ itself (any two vectors can be related by a linear transformation), but later we shall have to deal with bundles whose orbits are nontrivial and play a fundamental role. Since any orbit $S \subset F$ is by definition stable under the action of the structural group, if we choose any local trivialization $\bar{h}: \pi^{-1}(U) \to U \times F$ of the bundle, and define the subset $SU \subset \pi^{-1}(U)$ by

$$SU = \bar{h}^{-1}(U \times S)$$

this subset will be independent of the local trivialization chosen. This allows us to associate *locally* to each orbit $S \subset F$ a subset $SM \subset B$, which can be considered as a sub-bundle of $B$, with base $M$, fiber $S$ and structural group $G|S$ (the restriction of $G$ to the orbit $S$). In the examples we shall have to deal with, it will turn

---

†It is advisable to skip this part at first reading.

out that all the orbits $S$ are submanifolds of $F$, so that the $SM$ will also be submanifolds of $B$. Then it is only natural that the transverse space $T_b(B/SM)$ should play an important role. Also of importance will be the fact that "one looses nothing by restricting this transverse space to the fiber"—more precisely, let $F_x = \pi^{-1}(x)$ be the fiber above $x$, where $x = \pi(b)$, $b \in SM$. Then the inclusion mapping $F_x \subset B$ induces an isomorphism of the transverse spaces $T_b(F_x/S_x)$ and $T_b(B/SM)(S_x$ is $SM \cap F_x)$. Indeed, one has $T_b F_x \subset T_b B$, and also $T_b S_x \subset T_b(SM)$, so that this gives a linear mapping of the quotient spaces $T_b(F_x/S_x) \to T_b(B/SM)$. That this mapping is an isomorphism becomes obvious once the bundle has been trivialized in the neighborhood of $x$. Notice that although a trivialization is useful to establish the isomorphic character of the above mapping, the mapping itself is canonically defined, independently of any trivialization. It is important for the sequel to realize this canonical character of the isomorphism between the two transverse spaces. In the sequel, we shall often have to write that a section $\sigma : M \to B$ is transverse to the orbits of the bundle. According to Theorem 3, this condition is equivalent to the surjectivity of the mapping $T_x M \overset{T_x\sigma}{\to} T_x B \to T_b(B/SM)$, or, if one prefers, of the mapping $T_x M \to T_b(F_x/S_x)$ deduced from the former by the above isomorphism. In practice, the latter will be the only one we shall use, and we shall denote it by

$$\widetilde{T_x \sigma} : T_x M \to T_b(F_x/S_x)$$

bearing in mind that it is canonically deduced from $T_x \sigma$.


## 2.  ON SOME LOCAL SINGULARITIES OF MAPPINGS

Let $f : \mathbb{R}^m \to \mathbb{R}^n$ be a mapping of Euclidean spaces, with $m \geq n$. By a regular point of the mapping $f$ we shall mean a point of $\mathbb{R}^m$ where the Jacobian matrix of $f$ has rank $n$. We have seen in subsection 1.2 that near a regular point the mapping $f$ has the same local type as a canonical projection of Euclidean spaces (mappings having this local type have been called "submersions"). Now the question arises : what about the local type of $f$ near the "critical" points, the points which are not regular? This question is enormously difficult. In fact, Thom has shown that even for mappings defined by polynomials $(f_1, f_2, \ldots, f_n)$ there exists a nondenumerable infinity of local types.† This led him to restrict his attention to what he called "generic types," certain types which have some stability properties and are likely to be in finite number (for given $m$ and $n$). He then attempted to classify these generic types. Although Thom's classification of generic types is perhaps far from complete, it yields very interesting information, quite tractable too if one limits oneself to the simplest types of singularities. Here we shall make a thorough study of the very simplest type, called $S^1$ in the classification of Thom, which occurs "generically" for "almost all the critical points," and we shall also define (without describing them completely) the so-called $S^k$ types $(k = 2, 3, \ldots)$ which occur generically on critical sets of smaller dimension. Only the second derivatives of $f$ enter into the definition of these types. The more complicated types in Thom's classification [called $S^{k_1}(S^{k_2})$, $S^{k_1}(S^{k_2})$ $(S^{k_3})$], etc. involve the examination of higher-order derivatives of $f$, and we shall say nothing of them, since their very definition is sometimes problematic (it seems the problem has been solved very recently). Let us only stress the practical importance of the $S^1(S^1)$ type, or "Whitney cusp," which accounts for the frequent occurrence of cusps in many natural phenomena (cusps of caustics in geometrical optics, Riemann–Hugoniot phenomenon in fluid dynamics, cusps of Landau curves in the analytic study of Feynman integrals, etc.).

## 2.1. Preliminary Study of the Space of Linear Mappings

Let $L_m^n$ be the set of all linear mappings of the vector space $\mathbb{R}^m$ into the vector space $\mathbb{R}^n$. Obviously $L_m^n$ is an $mn$-dimensional manifold (isomorphic to $\mathbb{R}^{mn} =$ the set of all $m \times n$ matrices). There is a group acting naturally on $L_m^n$, namely the product $L(m) \times L(n)$ of the $m$-dimensional and the $n$-dimensional linear group. Its action is defined by multiplying the matrix $a \in L_m^n$ on the right by the matrix in $L(m)$ and on the left by the matrix in $L(n)$. Obviously the orbit of $a$ consists of those matrices having the same rank $r$ as $a$—because two matrices having the same rank can always be transformed one into the other by changing the bases in the vector

---

†In effect Thom dealt there only with the "topological types" (defined up to homeomorphisms) whereas, we are presently asking for a more refined classification; that of differential (or analytic) types. This is even worse.

spaces $\mathbb{R}^m, \mathbb{R}^n$. Such an orbit—the set of matrices whose rank is exactly $r$—will be denoted by $\overset{r}{S}$.

**Theorem 5.** $\overset{r}{S}$ *is a submanifold of $L^n_m$, of codimension $(m - r)$ $(n - r)$.*

The proof will be based on the following Lemma, which we leave to the reader as an easy exercise in linear algebra:

**Lemma.** *Let $a$ be an $m \times n$ matrix whose upper left $r \times r$ minor determinant $\Delta$ is non-zero. Then the rank of $a$ will be $r$ if and only if all $(r + 1) \times (r + 1)$ minors containing $\Delta$ have a vanishing determinant.*

In other words, to show that all the $(r + 1) \times (r + 1)$ minor determinants of $a$ vanish, it is sufficient to prove it for the minors $\Delta_{ij}$ obtained by adding to $\Delta$ a column in the $i$-th position $(i = r + 1, r + 2, \ldots, m)$ and a line in the $j$-th position $(j = r + 1, r + 2, \ldots, n)$.

Proof of Theorem 5: Let $a$ be a point of $\overset{r}{S}$, i.e., an $m \times n$ matrix of rank $r$. After eventually permuting the coordinates we can suppose that the upper left $r \times r$ minor determinant of $a$ does not vanish. Then the same property will, of course, also hold for all matrices sufficiently close to $a$. Therefore, applying the Lemma to these matrices, we see that $\overset{r}{S}$ coincides in the neighborhood of $a$ with the set of solutions of $(m - r)(n - r)$ equations, namely $\Delta_{ij} = 0$; $i = r + 1, r + 2, \ldots, m$; $j = r + 1, r + 2, \ldots, n$. The theorem will then follow from Theorem 2 if we show that the partial derivatives of the determinants $\Delta_{ij}$ with respect to all the matrix elements form an $mn \times (m - r)(n - r)$ matrix of maximum rank, but this is obvious since $(\partial \Delta_{i'j'}/\partial a_{ij})^{j,j'=r+1,\ldots,n}_{i,i'=r+1,\ldots,m}$ is just $\Delta$ times the unit matrix of $\mathbb{R}^{(m-r)(n-r)}$

The following theorem will prove very convenient to express transversality conditions with respect to $\overset{r}{S}$:

**Theorem 6.** *For all $a \in \overset{r}{S}$, the transverse space $T_a(L^n_m/\overset{r}{S})$ is canonically isomorphic to the space $L^{n-r}_{m-r}$ of linear mappings from the kernel $\mathrm{Ker}\, a$ to the cokernel $\mathrm{Coker}\, a$ (recall that the cokernel means the quotient vector space of the goal $\mathbb{R}^n$ by the image of the linear mapping $a$).*

Proof of Theorem 6: Consider the canonical inclusion $i_a$ : Ker $a \to \mathbb{R}^m$ and the canonical projection $p_a : \mathbb{R}^n \to$ Coker $a$. Define $\alpha_a : L_m^n \to L_{m-r}^{n-r}$ as the linear mapping which to every $b : \mathbb{R}^m \to \mathbb{R}^n$ associates

$$p_a \circ b \circ i_a : \text{Ker } a \to \text{Coker } a$$

Of course, $\alpha_a(a) = 0$. Now the tangent mapping to $\alpha_a$ at $a$ can be written

$$T_a \alpha_a : T_a L_m^n \to L_{m-r}^{n-r}$$

where we have used the fact that the manifold $L_{m-r}^{n-r}$ has a vector space structure, which enables us to identify it with its tangent space at the origin. Since the linear mapping $\alpha_a$ is surjective (as easily verified), $T_a \alpha_a$ also will be surjective. So if only we can prove that Ker $T_a \alpha_a = T_a \overset{r}{S}$, $T_a \alpha_a$ will induce the desired isomorphism between $T_a(L_m^n / \overset{r}{S})$ and $L_{m-r}^{n-r}$.

We shall prove it by two methods, the first of which is just an explicit verification with a special choice of coordinates, whereas, the second is more conceptual (but will not be formulated quite rigorously).

First Proof: Let us change the bases in $\mathbb{R}^m$ and $\mathbb{R}^n$ in such a way that the upper left $r \times r$ minor of $a$ becomes the unit matrix and all the other matrix elements vanish. Then Ker $a$ is the subspace of $\mathbb{R}^m$ generated by the $(r + 1, r + 2, \ldots, m)$ axes, whereas, Coker $a$ can be identified with the subspace of $\mathbb{R}^n$ generated by the $(r + 1, r + 2, \ldots, n)$ axes, and $\alpha_a$ is just the canonical projection $\mathbb{R}^{mn} \to \mathbb{R}^{(m-r)(n-r)}$ which to any $m \times n$ matrix $b$ associates its restriction to the lower right $(m - r) \times (n - r)$ rectangle. Thus, the Jacobian matrix of $\alpha_a$ is just

$$\frac{\partial [\alpha_a]_{i'j'}}{\partial b_{ij}} = \begin{cases} 1 & \text{if } (i, j) = (i', j') \\ 0 & \text{Otherwise} \end{cases}$$

which we find is equal to the Jacobian matrix $(\partial \Delta_{i'j'} / \partial a_{ij})(a)$ considered in Theorem 5.

Second Proof: Since $\overset{r}{S}$ is the orbit of $a$, $T_a \overset{r}{S}$ means intuitively the set of points in $L_m^n$ which are deduced from $a$ by the action of an infinitesimal element of the group $L(m) \times L(n)$, i.e., the set of matrices of the form $b = (1 + \eta) a (1 + \varepsilon)$, where $\varepsilon$ and $\eta$ are infinitesimal $m \times m$ and $n \times n$ matrices. Making $\alpha_a$ act on this, and retaining only terms of the first order in $(\varepsilon, \eta)$, we find

$$\alpha_a(b) = p_a(1 + \eta) \, a \, (1 + \varepsilon)i_a \simeq p_a \, a i_a + p_a \, a \varepsilon i_a + p_a \, \eta a i_a$$

but all these terms vanish, because $p_a \, a = 0$ and $a i_a = 0$.

This shows that $(T_a \alpha_a)(T_a \overset{r}{S}) = 0$, i.e., $T_a \overset{r}{S} \subset \text{Ker } T_a \alpha_a$. Then the strict equality of the two subspaces follows from the equality of their dimensions (admitting Theorem 5).

### 2.2.  A Bundle of Linear Mappings

We are now ready to start our study of the local singularities of the mapping $f : \mathbb{R}^m \to \mathbb{R}^n$ $(m \geq n)$. For conceptual reasons already explained at the beginning of Section 1.4, we prefer to consider $f$ as a mapping of manifolds

$$f : M \to N$$

A point $x \in M$ will be called critical of corank $k$ if the tangent mapping $T_x f : T_x M \to T_{f(x)} N$ has its corank at the goal equal to $k$. We would like to investigate how the corank varies when $x$ runs over M. For that purpose, let us consider the space $L_m^n(x)$ of all linear mappings from $T_x M$ to $T_{f(x)} N$. This is, of course, a manifold isomorphic to the manifold $L_m^n$ of Section 2.1, and the union $L_m^n M = \bigcup_{x \in M} L_m^n(x)$ can in an obvious fashion be made into a bundle with base $M$ and fiber $L_m^n$. In fact, any choice of local charts of $M$ and $N$ around $x$ and $f(x)$ allows us to identify $\bigcup_{x \in U} L_m^n(x)$ with the product $U \times L_m^n$, and any change of charts results in a linear transformation of $L_m^n$, namely, in multiplying he $m \times n$ matrix $a \in L_m^n$ on the right by a nonsingular $m \times m$ matrix and on the left by a nonsingular $n \times n$ matrix (the Jacobian matrices of the change of charts in $M$ and $N$)—this shows that the structural group of the bundle can be considered as the product of the $m$-dimensional and $n$-dimensional linear groups, acting on the fiber $L_m^n$ in the above fashion.

The orbits of the bundle $L_m^n M$ are the subsets $S^k M$ consisting of those linear mappings whose corank at the goal is $k : S^k M$ is a sub-bundle of $L_m^n M$, with fiber $S^k = \overset{n-k}{S} \subset L_m^n$.

Now by associating to every point $x \in M$ the tangent mapping $T_x f \in L_m^n(x)$ one defines a section of the bundle $L_m^n M$, which will be denoted by

$$\bar{f} : M \to L^n_m M$$

By definition, the point $x \in M$ will be critical of corank $k$ if and only if $\bar{f}(x) \in S^k$. We shall get further restriction on critical points by demanding that the section $\bar{f}$, which is obviously differentiable,† should be transverse to the orbits $S^k M \subset L^n_m M$ in the sense of Definition 11.

## 2. 3   Transversality Conditions Defining the $S^k$ Singularities

*First Transversality Condition.* A point $x \in M$ is called transversally critical of corank $k$ if $\bar{f}(x) \in S^k M$ and the section $\bar{f}$ is transverse to $S^k M$ at $x$.

*Theorem 7:   Near a transversally critical point of corank $k$, the set of critical points of corank $k$ is a submanifold $S^k_M \subset M$ of codimension $k(m - n + k)$—product of the coranks at the goal and at the source.*

This follows from Theorem 4, noticing that $S^k_M = \bar{f}^{-1}(S^k M)$. Notice that the dimension of the manifold $S^k_M$, as given by Theorem 7, is always smaller than the dimension $n$ of the goal. For instance,

$$\dim S^1_M = m - 1(m - n + 1) = n - 1$$

The mapping $f|S^k_M : S^k_M \to N$ is, therefore, a mapping from $S^k_M$ to a space of greater dimension, so that the simplest possibility for this mapping is to be an immersion.

*Second Transversality Condition.* Those points of $S^k_M$ where $f|S^k_M$ is an immersion are called *ordinary critical points of corank $k$* (the other points are called *exceptional*).

*Definition 12:*   By an $S^k$-type we shall mean the local type of a mapping $f$ near an ordinary critical point of corank $k$.

## 2.4.   Description of the $S^1$ Types.

First let us recall the local properties of $S^1$ types which are immediately deduced from the two transversality conditions.

---

†Differentiable of class $C^{r-1}$ if $f$ was differentiable of class $C^r$: In fact, the Jacobian matrix of $f$ in any system of coordinates will have a $C^{r-1}$ dependence on $x$.

First Transversality Condition $\Rightarrow$ *The critical set is a submani-fold $S_M^1 \subset M$ of dimension $n-1$.*

Second Transversality Condition $\Longleftrightarrow$ *The mapping $f|S_M^1 : S_M^1 \to N$ is an immersion.*

Since our study is local, i.e., we restrict ourselves to a small neighborhood of the critical point in $M$, we can suppose that $f|S_M^1$ is in effect an imbedding, and call $S_N^1$ the submanifold of codimension 1, isomorphic image of $S_M^1$ in $N$.

***Construction of a Local Model for f.*** In the goal $N$, one can choose local coordinates $(v_1, v_2, \ldots, v_{n-1}, y)$ such that $y$ is a local equation of the submanifold $S_N^1$. Then the functions $(v_1, v_2, \ldots, v_{n-1})$ can serve as local coordinates on the $(n-1)$ dimensional manifold $S_N^1$, and since $S_N^1$ is the isomorphic image of $S_M^1$ by $f$ the functions $u_i = v_i \circ f$ can serve as local coordinates on $S_M^1$. But since $S_M^1$ is a submanifold of $M$, the system $(u_1, u_2, \ldots, u_{n-1}, x_n, x_{n+1}, \ldots, x_m)$ where the $x_i$'s are local equations of $S_M^1$, will be a good local coordinate system on $M$. In these coordinates the mapping $f$ reads

$$v_1 = u_1$$
$$v_2 = u_2$$
$$\cdots\cdots$$
$$v_{n-1} = u_{n-1}$$
$$y = \varphi(u_1, u_2, \ldots, u_{n-1}, x_n, x_{n+1}, \ldots, x_m)$$

and the only problem is to give an explicit expression for the function $\varphi$. But considering the Jacobian matrix of $f$

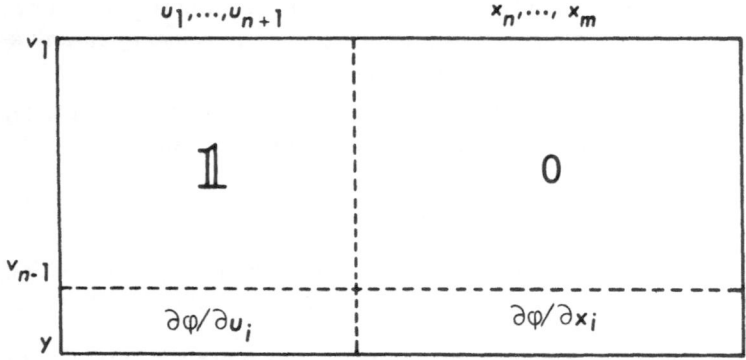

one sees that the critical set $S_M^1$ is given by the vanishing of all the $\partial\varphi/\partial x_i$. On the other hand, since $y$ vanishes on $S_N^1$, $\varphi$ must vanish on $S_M^1$. All this means that the function $\varphi$, as well as its first partial derivatives with respect to the $x_i$, must vanish for the zero values of all the arguments $x_i$. Therefore,† it can be put under the form

$$\varphi = \sum_{i,j=n,n+1,\ldots,m} x_i\, x_j\, a_{ij}(u_1, u_2, \ldots, u_{n-1}, x_n, x_{n+1}, \ldots, x_m) \quad (1)$$

To precise the properties of the matrix $(a_{ij})$, let us come back to the first transversality condition. We can consider $\bar{f}$ as the mapping of $M$ into $L_m^n$ which to every point of $M$, with coordinates $(u, x)$, associates the matrix just written a few lines above. We want to write that $\bar{f}$ is transverse at $(0, 0)$ to the submanifold $S^1 \subset L_m^n$ given locally (see Section 2.1) by the vanishing of the lower right rectangle drawn on the above matrix. According to Theorem 3, this is equivalent to saying that the mapping

$$M \to \mathbb{R}^{m-n+1}$$

$$(u, x) \mapsto (\partial\varphi/\partial x_i)$$

is a submersion. But since its tangent mapping at $(0, 0)$ just reads as the matrix

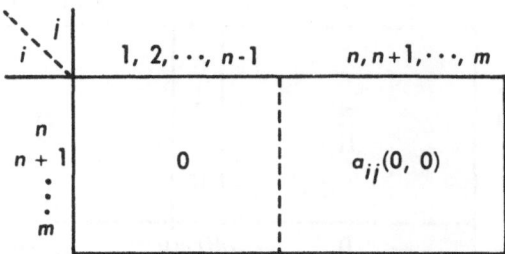

demanding it to be surjective just means one demands the square matrix $(a_{ij}(0, 0))$ to be nonsingular.

---

†For the reader who might think about the Taylor development of an analytic function, we give the more elementary proof, valid for differentiable functions as well. One just applies twice the following Lemma:

*If $\varphi(u, x)$ vanishes for the zero values of the $x_i$'s, then $\varphi(u, x) = \sum x_i\, a_i\, (u, x)$.*

To prove this lemma, one just notices that $\varphi(u, x) = \int_{t=0}^{t=1} d\varphi\ (u, tx)$

$= \int_0^1 \sum x_i(\partial\varphi/\partial x_i)(u, tx)dt$ so that, one can put

$$a_i(u, x) = \int_0^1 (\partial\varphi/\partial x_i)(u, tx)dt$$

Now comes the fundamental

*Morse Lemma: If a function $\varphi$ admits the form* (1), *with* $(a_{ij}(0))$ *a non-singular matrix, then after a suitable coordinate transformation*

$$(u_1, u_2, \ldots, u_{n-1}, x_n, x_{n+1}, \ldots, x_m) \mapsto (u_1, u_2, \ldots, u_{n-1}, x'_n, x'_{n+1}, \ldots, x'_m)$$

*it takes the form*

$$\varphi = \sum_{i=n, n+1, \ldots, m} \pm x'^2_i$$

*Proof:* The method is quite similar to the usual diagonalization of a bilinear form, except that we deal here with arbitrary coordinate transformations and not just linear ones, so that the inverse mapping theorem will have its role to play. The detail is as follows : After some permutation of the $x_i$, we can suppose that $a_{nn}(0) \neq 0$; putting $g(u, x) = \sqrt{|a_{nn}(u, x)|}$, we replace the coordinate $x_n$ by

$$x'_n = g(u, x) \, x_n + \sum_{j>n} \frac{a_{nj}(u, x)}{g(u, x)} \, x_j$$

This is a good local coordinate transformation, since its Jacobian matrix at the origin reads

and its determinant is $g(0, 0) \neq 0$. In these new coordinates $\varphi$ reads

$$\varphi = \pm x'^2_n + \varphi'$$

with

$$\varphi' = \sum_{j, i > n} x_i \, x_j \, a'_{ij}(u, x)$$

and one starts again with $\varphi'$ what has been done with $\varphi$, replacing $x_{n+1}$ by $x'_{n+1}$, etc.

All the discussion of this section is summarized in the following:

*Theorem 8: Near an ordinary critical point of corank 1 ($S_1$ type singularity), local coordinates $(x_1, x_2, \ldots, x_m)$ in the source, and $(y_1, y_2, \ldots, y_n)$ in the goal, can be chosen, such that the mapping reads*

$$y_1 = x_1$$
$$y_2 = x_2$$
$$\cdots\cdots$$
$$y_{n-1} = x_{n-1}$$
$$y_n = \pm x_n^2 \pm x_{n+1}^2 \pm \cdots \pm x_m^2$$

The above relations will be called the *canonical model of an $S^1$ type singularity*. The number of minus signs in the latter quadratic form is called the *transverse index of the critical point*. The reason for this denomination is that, as we shall see, this quadratic form can be given an intrinsic meaning (up to multiplication by a scalar factor) as a quadratic form on the space $T_x(M/S_M^1)$ of transverse vectors to $S_M^1$. This quadratic form (or the corresponding bilinear form) will be called the "transverse Hessian"[*] of the critical point. The "transverse index" is just the index of the transverse Hessian, i.e., the greatest possible dimension of a subspace on which this quadratic form is negative definite. Since the transverse Hessian will be given a completely intrinsic definition, the transverse index has an intrinsic meaning (i.e., one cannot write for the same mapping $f$ two canonical models with different transverse indices) at least modulo the change $i \to (m - n + 1) - i$ [because the transverse Hessian will be defined only up to multiplication by a scalar, and if this scalar is negative the index $i$ is changed into its complement $(m + n - 1) - i$].

In the next section we shall show how to construct the transverse Hessian, but instead of working with the quotient space $T_x(M/S_M^1) = T_x(M)/T_x S_M^1$ we shall work with the subspace Ker $T_x f \subset T_x M$, which is clearly isomorphic to the former in the case of an $S^1$ singularity. The reason for preferring the latter is that it is defined not only for $S^1$ singularities but for an arbitrary critical point. The transverse Hessian will also be constructed for such an arbitrary critical point (of corank 1), and we shall see that the $S^1$ character of the critical point is equivalent to the non-degeneracy of the transverse Hessian.

## 2.5. Reformulation of the Transversality Conditions.  Transverse Hessian of a Critical Point of Corank 1.

As already noticed at the end of Section 1.5, the tangent mapping $T_x \bar{f} : T_x M \to T_{\bar{f}(x)}(L_m^n M)$ to the section $\bar{f} : M \to L_m^n M$ determines canonically a linear mapping $\widetilde{T_x \bar{f}} : T_x M \to T_{f(x)}(L_m^n(x)/S^k(x))$, where $L_m^n(x)$ is the space of linear mappings from $T_x M$ to $T_{f(x)}N$ (fibre of $L_m^n M$), and $S^k(x)$ is the subspace of those mappings which have the same corank $k$ as $\bar{f}(x)$.  The first transversality condition is obviously equivalent to the surjectivity of the above linear mapping :

*First Transversality Condition* $\Longleftrightarrow$ $\widetilde{T_x \bar{f}}$ *is Surjective.*  Now what about the second transversality condition?  It asks $f|S_M^k : S_M^k \to N$ to be an immersion, i.e., $T_x f|T_x S_M^k : T_x S_M^k \to T_{f(x)}N$ to be injective.  In other words, a vector of $T_x S_M^k$ must be zero if its image is zero.  But since $S_M^k = \bar{f}^{-1}(S^k)$, it follows from the first transversality condition that $T_x S_M^k = (T_x \bar{f})^{-1}(T_{f(x)}S^k) = (\widetilde{T_x \bar{f}})^{-1}(0)$ $= \mathrm{Ker}\ \widetilde{T_x \bar{f}}$.  Therefore, the second transversality condition is equivalent to the condition

$$\mathrm{Ker}\ \widetilde{T_x \bar{f}} \cap \mathrm{Ker}\ T_x f = (0)$$

which can also be conveniently expressed as an injectivity condition for the restricted mapping

$$\widehat{T_x \bar{f}} = \widetilde{T_x \bar{f}}|\mathrm{Ker}\ T_x f :$$

*Second Transversality Condition* $\Longleftrightarrow$ $\widehat{T_x \bar{f}}$ *is Injective.*  The advantage of this new formulation is that it allows to enounce each transversality condition independently from the other, whereas previously the formulation of the second condition was possible only if the first was satisfied.  This is a big advantage in the case of corank 1.  In fact, when $k = 1$, $\widehat{T_x \bar{f}}$ is a linear mapping from the $(m - n + 1)$—dimensional vector space $\mathrm{Ker}\ T_x f$ to a space of the same dimension.  Therefore, saying that $\widehat{T_x \bar{f}}$ is injective is equivalent to saying it is surjective (or bijective, as you like).  But since $\widehat{T_x \bar{f}}$ is a restriction of $\widetilde{T_x \bar{f}}$, the surjectivity of the former implies the surjectivity of the latter, so that the second transversality condition implies the first.  This leads to the following

*Theorem 9:  The $S^1$ character of a critical point of corank 1 is equivalent to the nondegeneracy of the linear mapping*

$$\widehat{T_x \bar{f}} : \mathrm{Ker}\ T_x f \to T_{f(x)}(L_m^n(x)/S^1(x))$$

But remember that Theorem 6 gave us a canonical identification between $T_{\bar{f}(x)}(L_m^n(x)/S^k(x))$ and the space $L_{m-n+k}^k(x)$ of linear mappings from Ker $\bar{f}(x)$ to Coker $\bar{f}(x)$. In the $k = 1$ case, we can identify the one-dimensional vector space Coker $\bar{f}(x)$ with the real line $\mathbb{R}$, so that $L_{m-n+1}^1(x)$ is identified with the dual space [Ker $\bar{f}(x)]^*$. In this way, remembering that $\bar{f}(x) = T_x f$, we see that $\widehat{T_x \bar{f}}$ can be identified with a linear mapping

$$\widehat{T_x \bar{f}} : \text{Ker } T_x f \to [\text{Ker } T_x f]^*$$

A linear mapping from a vector space into its dual is a *bilinear form*; we have thus identified $\widehat{T_x \bar{f}}$ with a bilinear form on Ker $T_x f$: *this bilinear form is the transverse Hessian*. The only arbitrary step in the construction has been the identification of the one dimensional vector space Coker $T_x f$ with the real line $\mathbb{R}$. It is just a "scale arbitrariness," which results in the transverse Hessian being defined only up to a nonvanishing scalar factor.

### 2.6. A Practical way of Verifying the $S^1$ Character of a Critical Point

This subsection is just the analytical translation, in arbitrary coordinates, of the geometrical concepts we have just discussed. Let $f$ be given in terms of local coordinates by $n$ functions $f_1, f_2, \ldots, f_n$ of variables $x_1, x_2, \ldots, x_m$. Then $\bar{f}$ can be identified with the mapping $\bar{f} : M \to L_m^n$ which to any point $x \in M$ associates the matrix $(a_i^j(x) = (\partial f_j/\partial x_i)(x))_{i,j}$, so that $T_x \bar{f} : T_x M \to T_{\bar{f}(x)} L_m^n$ is the mapping which to any tangent vector $X$ at $x$ associates the tangent vector $A(X)$ given by

$$A_i^j(X) = \sum_k \frac{\partial a_i^j(x)}{\partial x_k} X_k = \sum_k \frac{\partial^2 f_j(x)}{\partial x_i \partial x_k} X_k$$

Now let $\bar{x}$ be a critical point of corank 1, and consider the associated matrix $\bar{a} = (\bar{a}_i^j = (\partial f_j/\partial x_i)(\bar{x}))_{i,j}$. Saying that $\bar{x}$ is critical of corank 1 is equivalent to saying that there exist scalar parameters $\alpha_1, \alpha_2, \ldots, \alpha_n$ not all vanishing, well defined up to a common multiplicative factor, such that

$$\sum_{j=1}^n \alpha_j \bar{a}_i^j = 0 \qquad i = 1, 2, \ldots, m$$

(This condition indeed expresses the linear dependence of the lines of the matrix $\bar{a}$). Thus, denoting by $\alpha : \mathbb{R}^n \to \mathbb{R}$ the surjective linear mapping which to any vector $(a^j) \in \mathbb{R}^n$ associates the number $\sum_j \alpha_j a^j$, we see that $\alpha \circ \bar{a} = 0$, so that one can identify $\alpha$ with the canonical projection of $\mathbb{R}^n$ on Coker $\bar{a}$, identifying $\mathbb{R}$ with that cokernel. With this identification, the mapping

$$\alpha_a : L_m^n \to L_{m-n+1}^1$$

considered in Theorem 6 is just the mapping which to any matrix $(a_i^j) \in L_m^n$ associates the $m$-uple of numbers $\sum_{j=1}^n \alpha_j a_i^j$ considered as a linear mapping from Ker $\bar{a} \subset \mathbb{R}^m$ to $\mathbb{R}$. Similarly, the tangent mapping

$$T_a \alpha_a : T_a L_m^n \to T_0 L_{m-n+1}^1 = L_{m-n+1}^1$$

associates to the matrix $(A_i^j)$ the $m$-uple of numbers $\sum_{j=1}^m \alpha_j A_i^j$ considered as a linear mapping from Ker $\bar{a}$ to $\mathbb{R}$.

Putting all this together, we see that $\widehat{T_x f}$ *is the bilinear form which to any couple of vectors* $X$, $Y$ *in* Ker $T_x f$ *associates the number*

$$\sum_{j=1}^n \sum_{i,k=1}^m \alpha_j \frac{\partial^2 f(\bar{x})}{\partial x_i \partial x_k} X_i Y_k$$

*The condition that* $X$ *must belong to* Ker $T_x f$ *reads*

$$\sum_{i=1}^m \frac{\partial f_j(\bar{x})}{\partial x_i} X_i = 0 \qquad j = 1, 2, \ldots n$$

(and similarly for $Y$). Therefore, the degeneracy of the bilinear form $\widehat{T_x f}$ (i.e., the non $S^1$ character of the critical point $x$) is equivalent to the existence of a vector $X \in$ Ker $T_x f$ such that the vector $(\sum_{j=1}^n \sum_{i=1}^m \alpha_j (\partial^2 f_j(\bar{x})/\partial x_i \partial x_k) X_i)_k$, considered as an element of the dual of Ker $T_x f$, vanishes. More explicitly, we can enounce:

*Theorem 10:* A critical point $\bar{x}$ of corank 1 will *not* be of type $S^1$ if and only if there exist an $m$-uple of numbers $(X_i)$, and an $n$-uple of numbers $(\beta_j)$, satisfying the equations

$$\sum_{i,j} \alpha_j \frac{\partial^2 f_j(\bar{x})}{\partial x_i \partial x_k} X_i + \sum_j \beta_j \frac{\partial f_j(\bar{x})}{\partial x_k} = 0 \qquad k = 1, 2, \ldots, m$$

$$\sum_i \frac{\partial f_j(\bar{x})}{\partial x_i} X_i = 0 \qquad j = 1, 2, \ldots, n$$

where $(\alpha_j)$ is the $n$-uple of numbers, well defined up to a common multiplicative factor, solution of the equations

$$\sum_j \alpha_j \frac{\partial f_j(\bar{x})}{\partial x_i} = 0 \qquad i = 1, 2, \ldots, m$$

### 2.7. A Last Remark

The main purpose of this final subsection is to prove the following:

*Theorem 11:*  Consider the following commutative diagram of mappings of manifolds

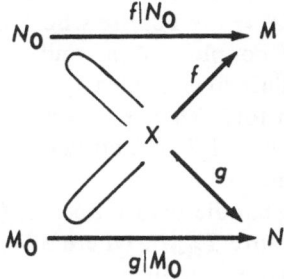

where $f$ and $g$ are submersions and $M_0 = f^{-1}(0)$, $N_0 = g^{-1}(0)$. *Then, at a point $x \in M_0 \cap N_0$, $g|M_0$ is of type $S^1$ if and only if $f|N_0$ is of type $S^1$, and in that case, the transverse indices of these two mappings are the same.*

Before starting the proof, let us explain the usefulness of this theorem. One is often faced with such a situation: $g: \mathbb{R}^l \to \mathbb{R}^n$ is a canonical projection of Euclidean spaces, and one wants to study the singularities of this projection restricted to some submanifold $M_0 \subset \mathbb{R}^l$; it may be rather inconvenient to write the explicit expression of the mapping $g|M_0$ in terms of local coordinates on $M_0$. One would much more prefer to work directly on the local equations $f_1, f_2, \ldots, f_m$ of the submanifold $M_0$. The above theorem enables us to do so, showing that $g|M_0$ will be of type $S^1$ if and only if these local equations, restricted to the fiber $\mathbb{R}^{l-n} = g^{-1}(0)$, constitute a mapping of type $S^1$.

A very natural way to prove the theorem is to notice that the tangent mappings $T_x(f|N_0)$ and $T_x(g|M_0)$ have the same kernel and the same cokernel. This shows that the coranks at the source—and at the goal—of these two mappings are the same. Then one considers, on the common kernel, the bilinear forms defined by the transverse Hessians of these two mappings; one verifies that these bilinear forms coincide, and that proves the theorem.

The method we shall choose to verify this identity of the transverse Hessians is rather lengthy when written in full detail, but it will look very natural to the topologically minded reader, and it will give us precious information also on $S^k$-type singularities for $k > 1$. The basic idea is that everything should be re-expressed in terms of the relative position of the two subspaces $T_x M_0$, $T_x N_0$ in $T_x X$.

WE SHALL INTRODUCE THE FOLLOWING BUNDLES ABOVE $X$:

(1)  $(G_{m_0}^m \times G_{n_0}^n)X$ = the bundle whose fiber $(G_{m_0}^m \times G_{n_0}^n)(x)$ consists of all possible couples of $m_0$ and $n_0$-dimensional subspaces $(\prod_{m_0}, \prod_{n_0})$ of $T_x X$. This fiber is clearly a manifold, the product of two Grassmann manifolds (hence, the notation). The group of the bundle is the *linear group* of $T_x X$, acting on the Grassmann manifolds in an obvious fashion.

(2)  $L_{m_0}^n X$ = the bundle whose fiber $L_{m_0}^n(x)$ consists of all linear mappings of $T_x M_{f(x)}$ into $T_{g(x)} N$, where $M_{f(x)}$ stands for $f^{-1}(f(x))$.

(3)  $L_{n_0}^m X$ analogously defined.

WE SHALL CONSIDER THE FOLLOWING SECTIONS OF THESE BUNDLES:

(1)  $(\overline{f, g}): X \to (G_{m_0}^m \times G_{n_0}^n)X$ which to $x$ associates the couple $(T_x M_{f(x)}, T_x N_{g(x)})$;

(2)  $\bar{g}_M: X \to L_{m_0}^n X$ which to $x$ associates the linear mapping $T_x(g|M_{f(x)}): T_x M_{f(x)} \to T_{g(x)} N$

(3)  $\bar{f}_N: X \to L_{n_0}^m X$ analogously defined.

THE ORBITS OF THESE BUNDLES ARE THE FOLLOWING SUB-BUNDLES:

(1)  ${}^k\sum X \subset (G_{m_0}^m \times G_{n_0}^n)X$, consisting of those couples $(\prod_{m_0}, \prod_{n_0})$ such that $\prod_{m_0} \cap \prod_{n_0}$ has codimension $m + n - k$ in $T_x X$;

(2)  ${}^k S_{m_0}^n X \subset L_{m_0}^n X$, consisting of those mappings whose corank at the goal is exactly $k$;

(3)  ${}^k S_{n_0}^m X \subset L_{n_0}^m X$, similarly defined.

WE SHALL THEN RELATE ALL THESE BUNDLES BY CONSTRUCTING "FIBER MAPPINGS"† $\varphi$ AND $\gamma$

$$L_{n_0}^m X \xleftarrow{\varphi} (G_{m_0}^m \times G_{n_0}^n)X \xrightarrow{\gamma} L_{m_0}^n X$$

---

†"Fiber mapping" means that the fiber above any point $x$ is sent into the fiber above the same point $x$.

ENDOWED WITH THE FOLLOWING FUNDAMENTAL PROPERTIES:

(*P 1*)  *relating the sections:* The diagram

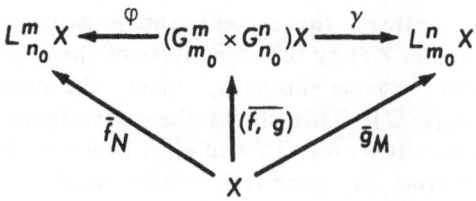

is commutative;

(*P 2*)  *relating the orbits:*

$$^k\textstyle\sum X = \gamma^{-1}(^k S_{m_0}^n X) = \varphi^{-1}(^k S_{n_0}^m X)$$

(*P 3*)  *a transversality condition:* $\gamma$ is transverse to $^k S_{m_0}^n X$, and $\varphi$ is transverse to $^k S_{n_0}^m X$;

(*P 4*)  *demanding the commutativity of the following diagram:*

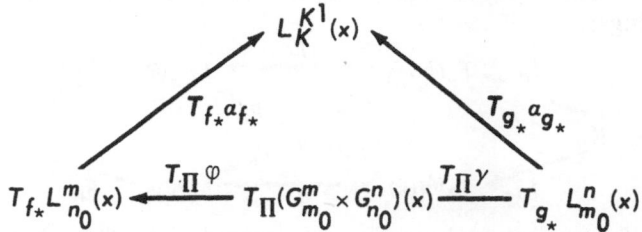

where $L_K^{K'}(x)$ is the space of linear mappings from the common kernel $K$ to the common cokernel $K'$ of $T_x(g|M_0)$ and $T_x(f|N_0)$, $(x \in M_0 \cap N_0)$; $\prod = (\bar{f}, \bar{g})(x)$; $f_* = \varphi(\prod) = \bar{f}_N(x)$; $g_* = \gamma(\prod) = \bar{g}_M(x)$; $\alpha_{f_*}$ (and similarly $\alpha_{g_*}$) is the linear mapping defined in the proof of Theorem 6.

Properties (P 2) and (P 3) imply that the orbits $^k\textstyle\sum X$ are submanifolds (Theorem 4), and that the tangent mapping to $\varphi$ (respectively, $\gamma$) induces an *isomorphism between the transverse space to* $^k\textstyle\sum X$ *and the transverse space to* $^k S_{n_0}^m X$ (respectively, $^k S_{m_0}^n X$). On the other hand, we already know, by Theorem 6, that the tangent mapping to $\alpha_{f_*}$ (respectively, $\alpha_{g_*}$) induces an *isomorphism between the transverse space to* $^k S_{n_0}^m(x)$ (respectively, $^k S_{m_0}^n$) *and* $L_K^{K'}(x)$.

*Property* (P 4) *shows that all the above isomorphisms are compatible.* This, together with property (P 1), shows that *the sections* $\bar{f}_N$ *and* $\bar{g}_M$ *induce the same linear mapping from* $T_x X$ *to* $L_K^{K'}(x)$.

Now, the first transversality condition for $f|N$ (respectively, $g|M$) is expressed by the surjectivity of the above linear mapping restricted to $T_xN$ (respectively, $T_xM$): the source is *not the same* for $f|N$ and $g|M$. On the contrary, the second transversality condition is expressed (see section 2.5) by the injectivity of the above linear mapping restricted to the *same* subspace, namely the common kernel of $T_x(f|N)$ and $T_x(g|M)$. This shows the equivalence of the second transversality condition for $f|N$ and $g|M$. Since in the corank 1 case the second transversality condition implies the first we thus get the equivalence of the $S^1$ characters of $f|N$ and $g|M$. Furthermore, since the above restricted linear mapping is what has been called the "transverse Hessian" (Section 2.5), this transverse Hessian (and, therefore, the transverse index) is the same for $f|N$ and $g|M$, and Theorem 11 is completely proved.†

*Construction of the Fiber Mapping* $\gamma$. The various tangent spaces which will play a role in the construction can be conveniently arranged into the following diagram of vector spaces and linear mappings:

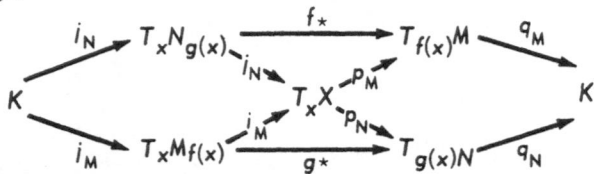

where for brevity one has written $f_*$ instead of $T_x(f|N_{g(x)})$, $g_*$ instead of $T_x(g|M_{f(x)})$, $p_M$ instead of $T_xf$, $p_N$ instead of $T_xg$ (notice that $T_{f(x)}M$ can be canonically identified with the quotient space $T_xX/T_xM_{f(x)}$, and $p_M$ with the canonical projection; similarly, $T_{g(x)}N$ can be canonically identified with $T_xX/T_xN_{g(x)}$, and $p_N$ with the canonical projection).

We also have represented on the diagram the common kernel $K$ of $f_*$ and $g_*$; this is nothing but $T_xM_{f(x)} \cap T_xN_{g(x)}$. Also $K'$ denotes the common cokernel of $f_*$ and $g_*$, which is nothing but $T_xX/(T_xM_{f(x)} + T_xN_{g(x)})$. All the letters $i, j$ on the diagram represent canonical injections of vector subspaces, whereas, the letters $p, q$

---

†Notice that to establish the equivalence of the $S^1$ characters we only needed the isomorphism of the transverse spaces, which followed from properties (P 2) and (P 3). Property (P 4) was only needed to prove the equality of the transverse Hessians, used to establish the equality of the transverse indices.

represent canonical projections from a vector space to a quotient space.

As already mentioned the coincidence of the kernels (respectively, cokernels) of $f_*$ and $g_*$ implies that these two mappings have the same corank at the source (respectively, at the goal). Moreover the dimension $\kappa$ of the kernel $K = T_x M_{f(x)} \cap T_x N_{g(x)}$ is related to the corank at the goal $k$ by the formula

$$l - \kappa = m + n - k$$

where $l$ is the dimension of $X$.

A good way of guessing how to construct the mapping $\gamma$ is to look at the following canonical mappings:

$$(G_{m_0}^m \times G_{n_0}^n)(x) \xleftarrow{\text{Im} \times \text{Ker}} (L_{m_0}^l \times L_l^n)(x) \xrightarrow{\circ} L_{m_0}^n(x) \qquad (2)$$

where $\circ$ is the "composition mapping," which to any couple of linear mappings $\alpha: T_x M_{f(x)} \to T_x X$, $\beta: T_x X \to T_{g(x)} N$ associates the composed mapping $\beta \circ \alpha$; $Im$ (respectively, Ker) is the mapping which to $\alpha$ (respectively, $\beta$) associates the subspace Im $\alpha$ (respectively, Ker $\beta$) $\subset T_x X$: this is indeed an $m_0$ (respectively, $n_0$)-dimensional subspace, provided $\alpha$ (respectively, $\beta$) is sufficiently close to $i_M$ (respectively, $p_N$) since then $\alpha$ (respectively, $\beta$) will be an injection (respectively, a surjection). Now one looks at the above two canonical mappings and one thinks that if it were possible to "reverse" the left arrow one could define $\gamma$ by just composing the two arrows. Since this left arrow Im $\times$ Ker is obviously surjective, the most natural way to reverse it is to construct a "section" of it (recall that a section $s: A \to B$ of a surjection $p: B \to A$ is a mapping such that $p \circ s = 1_A$).

*Construction of a Section of* Im $\times$ Ker. One starts by choosing sections $s_M$ and $s_N$ of the projections $p_M$, $p_N$ (see the diagram on p. 96), this for every value of $x$. When $x$ runs in a small domain in $X$, it is easy to make this choice depend continuously on $x$. Notice that the choice of a section of $p_M$ (respectively, $p_N$) amounts to identifying the quotient space $T_{f(x)} M$ (respectively, $T_{g(x)} N$) with a subspace of $T_x X$. One then can identify $T_x X$ with the direct sum $T_x M_{f(x)} \oplus T_{f(x)} M$ (respectively, $T_x N_{g(x)} \oplus T_{g(x)} N$): this is done formally by the following diagram:

$$T_x M_{f(x)} \underset{r_M}{\overset{i_M}{\rightleftarrows}} T_x X \underset{s_M}{\overset{p_M}{\rightleftarrows}} T_{f(x)} M$$

respectively,

$$T_x N_{g(x)} \underset{r_N}{\overset{i_N}{\rightleftharpoons}} T_x X \underset{s_N}{\overset{p_N}{\rightleftharpoons}} T_{g(x)} N$$

satisfying

1) $p \circ s = 1$                  (meaning that $s$ is a "section" of $p$)
2) $r \circ i = 1$                      ($r$ is a "retraction")
3) $s \circ p + i \circ r = 1$

(Exercise: show that given $s$, there exists one and only one $r$ satisfying these properties).

Now, as in our description of Grassmann manifolds (section 1.3), we can identify some domain of $(G^m_{m_0} \times G^n_{n_0})(x)$ with $(L^m_{m_0} \times L^n_{n_0})(x)$, where $L^m_{m_0}(x)$ (respectively, $L^n_{n_0}(x)$) denotes the set of linear mappings of $T_x M_{f(x)}$ into $T_{f(x)} M$ (respectively, $T_x N_{g(x)}$ into $T_{g(x)} N$.) In fact, any subspace $\prod_{m_0}$ transverse to the subspace $s_M(T_{f(x)} M)$ in $T_x X$ can be considered as the graph of a linear mapping from $T_x M_{f(x)}$ to $T_{f(x)} M$. Therefore, *everywhere in the following, we shall replace* $(G^m_{m_0} \times G^n_{n_0})X$ *by* $(L^m_{m_0} \times L^n_{n_0})X$. The section $(\overline{f, g})$ will then be replaced by *the zero section* of $(L^m_{m_0} \times L^n_{n_0})X$, which to any $x$ associates the zero mappings. The diagram (2) is replaced by

$$(L^m_{m_0} \times L^n_{n_0})(x) \xoverset{(p_M \cdot) \times (\cdot i_N)}{\longleftarrow} (L^l_{m_0} \times L^n_l)(x) \overset{\circ}{\longrightarrow} L^n_{m_0}(x)$$

where $(p_M \cdot)$ means the mapping which to any $\alpha : T_x M_{f(x)} \to T_x X$ associates $p_M \circ \alpha : T_x M_{f(x)} \to T_{f(x)} M$, and $(\cdot i_N)$ means the mapping which to any $\beta : T_x X \to T_{g(x)} N$ associates $\beta \circ i_N : T_x N_{g(x)} \to T_{g(x)} N$ [i.e., the restriction of $\beta$ to $T_x N_{g(x)}$].

Now it is easy to construct sections of $(p_M \cdot)$ and $(\cdot i_N)$,

$$L^m_{m_0}(x) \underset{\tilde{s}_M}{\overset{(p_M \cdot)}{\rightleftharpoons}} L^l_{m_0}(x)$$

$$L^n_{n_0}(x) \underset{\tilde{r}_N}{\overset{(\cdot i_N)}{\rightleftharpoons}} L^n_l(x)$$

by simply associating, to each $a \in L^m_{m_0}(x)$, $\tilde{s}_M(a) = i_M + s_M \circ a$ and to each $b \in L^n_{n_0}(x)$, $\tilde{r}_N(b) = p_N + b \circ r_N$ (Exercise: verify that $\tilde{s}_M$ and $\tilde{r}_N$ are sections). The two diagrams below are to help the reader visualizing these definitions

$$T_x M_{f(x)} \overset{i_M}{\longrightarrow} T_x X \overset{s_M}{\longleftarrow} T_{f(x)} M$$
$$a$$

$$T_x N_{g(x)} \overset{r_N}{\longleftarrow} T_x X \overset{p_N}{\longrightarrow} T_{g(x)} N$$
$$b$$

*Now everything is ready, and we can define $\gamma$ as the composed mapping*

$$\gamma: (L^m_{m_0} \times L^n_{n_0})(x) \xrightarrow{\tilde{s}_M \times \tilde{r}_N} (L^l_{m_0} \times L^n_l)(x) \xrightarrow{\circ} L^n_{m_0}(x)$$

Explicitly, $\gamma$ transforms the couple $(a, b)$ into

$$\gamma(a, b) = (p_N + b \circ r_N) \circ (i_M + s_M \circ a)$$
$$= g_* + p_N \circ s_M \circ a + b \circ r_N \circ i_M + b \circ r_N \circ s_M \circ a$$

Notice that $\gamma(0, 0) = g_*$, so that $\gamma$ *transforms the zero section into the section $\tilde{g}_M$, in conformity with property* (P 1).

*Let us verify now property* (P 2). Since $\tilde{s}_M \times \tilde{r}_N$ is a section of Im $\times$ Ker, the reciprocal image $\gamma^{-1}({}^k S^n_{m_0})$ is nothing but the projection (through Im $\times$ Ker) of the reciprocal image $\circ^{-1}({}^k S^n_{m_0})$: the latter consists of the couples $(\alpha, \beta) \in L^l_{m_0} \times L^n_l$ such that $\beta \circ \alpha$ has corank $k$ at the goal, i.e., corank $m_0 - n + k$ at the source, i.e., its kernel is $(m_0 - n + k)$-dimensional; but since $\alpha$ is injective, Ker $(\beta \circ \alpha)$ is isomorphic with Im $\alpha \cap$ Ker $\beta$ so the statement that $(\alpha, \beta) \in \circ^{-1}({}^k S^n_{m_0})$ is equivalent to saying that the image of $(\alpha, \beta)$ through the mapping Im $\times$ Ker is $(m_0 - n + k)$-dimensional, i.e., belongs to ${}^k\Sigma$, in conformity with property (P 2).

*Next we must verify property* (P 3). According to Theorem 3 and 6, saying that $\gamma$ is transverse to ${}^k S^n_{m_0}$ is equivalent to saying that the composed mapping $\alpha_{g_*} \circ \gamma$ is a *submersion*, where

$$\alpha_{g_*}: L^n_{m_0}(x) \longrightarrow L^{K'}_K$$

is defined by $\alpha_{g_*}(c) = q_N \circ c \circ j_M$ ($L^{K'}_K$ stands for the set of linear mappings from the kernel $K$ to the cokernel $K'$ of $g_*$).

Let us write explicitly what $\alpha_{g_*} \circ \gamma$ is. It transforms the couple $(a, b) \in (L^m_{m_0} \times L^n_{n_0})(x)$ into

$$(\alpha_{g_*} \circ \gamma)(a, b) = q_N \circ p_N \circ s_M \circ a \circ j_M + q_N \circ b \circ r_N \circ i_M \circ j_M$$
$$+ q_N \circ b \circ r_N \circ s_M \circ a \circ j_M$$
$$= q_M \circ a \circ j_M + q_N \circ b \circ j_N + q_N \circ b \circ r_N \circ s_M \circ a \circ j_M$$

where we have used the facts that $q_N \circ p_N = q_M \circ p_M$, $i_M \circ j_M = i_N \circ j_N$, $p_M \circ s_M = 1$ and $r_N \circ i_N = 1$. Now we want to show that near the point $(0, 0)$ the mapping $\alpha_{g_*} \circ \gamma$ is a submersion. We show below that $\alpha_{g_*} \circ \gamma$ is surjective separately in each of its variables when the other is fixed at the value zero, a fact which suffices to establish the submersive character in view of the bilinearity of $\alpha_{g_*} \circ \gamma$ (a surjective linear mapping is a submersion). And indeed, let $e$ be any element of $L^{K'}_K$. Choose any section $\sigma_M$ of the projection $q_M$ and any retraction

$\rho_M$ of the inclusion $j_M$. Then $\sigma_M \circ e \circ \rho_M \in L_{m_0}^m(x)$, and $(\alpha_{g_{\bullet}} \circ \gamma)(\sigma_M \circ e \circ \rho_M, 0)$ $= q_M \circ \sigma_M \circ e \circ \rho_M \circ j_M = e$, thus proving the surjectivity.

*There remains to verify property* (P 4), i.e., the equality of the tangent mappings to $\alpha_{g_{\bullet}} \circ \gamma$ and $\alpha_{f_{\bullet}} \circ \varphi$, considered at the point $\prod =$ $(\overline{f}, \overline{g})(x)$. Identifying as previously $(G_{m_0}^m \times G_{n_0}^n)(x)$ with $(L_{m_0}^m \times L_{n_0}^n)(x)$, we have already given an expression for $\alpha_{g_{\bullet}} \circ \gamma$, namely,

$$(\alpha_{g_{\bullet}} \circ \gamma)(a, b) = q_M \circ a \circ j_M + q_N \circ b \circ j_N + q_N \circ b \circ r_N \circ s_M \circ a \circ j_M$$

The fiber mapping $\varphi$ can be constructed in complete analogy with $\gamma$, just exchanging the roles of $M$ and $N$, $f$ and $g$, etc. we shall thus get

$$(\alpha_{f_{\bullet}} \circ \varphi)(a, b) = q_N \circ b \circ j_N + q_M \circ a \circ j_M + q_M \circ a \circ r_M \circ s_N \circ b \circ j_N$$

We see that these two mappings $(\alpha_{g_{\bullet}} \circ \gamma)$ and $(\alpha_{f_{\bullet}} \circ \varphi)$ coincide to the first order in $(a, b)$. This means that their tangent mappings at the origin $(0, 0)$ coincide. But this is just Property (P 4), since in our construction the zero section represents the section $(\overline{f}, \overline{g})$.


## BIBLIOGRAPHIC ANNOTATION

An exposition of the basic concepts of differential topology can be found in the beginning of the book of S. Sternberg, "Lectures on Differential Geometry," Prentice-Hall Mathematics Series, Prentice Hall, (1964).

Concerning the nondenumerable infinity of topological types of polynomial mappings, see R. Thom, "L'Enseignement mathématique," **VIII**: 24–33 (1962).

For the notion of "generic type," and the definition of $S^k$ types, $S^k(S^{k'})$ types, etc., see R. Thom, *Ann. Inst. Fourier*, **6**: 43–87 (1956).

The special case when the goal is one-dimensional constitutes "Morse Theory" on which a very illuminating book is available: J. Milnor, "Morse Theory," (*Ann. of Math.* Studies No. 51) Princeton University Press, (1963).

The mappings of the plane into the plane are investigated in detail by H. Whitney, *Ann. of Math.* **62** (3): 374–410 (1965).

# Formal Aspects of Potential Theory

## M. SCADRON

*IMPERIAL COLLEGE OF SCIENCE AND TECHNOLOGY†*
*London, England*

---

## 1. INTRODUCTION

Recent years have seen new approaches toward the mathematical problems of scattering theory. In what follows, we will summarize some of the formal techniques that have been applied to the non-relativistic scattering of a particle in a fixed source potential $V$. Hopefully, one or more of these ideas may some day prove useful for the understanding of the more complicated relativistic and $n$-body theories.

To begin with, we start with the well-known Lippmann–Schwinger (L–S) equation for scattering states.

$$|\psi(s)_+\rangle = |\phi(s)\rangle + G_0(s)V|\psi(s)_+\rangle \tag{1}$$

where $G_0(s) = (s - H_0 + i\varepsilon)^{-1}$, $s$ being the energy $k^2$ and $|\phi(s)\rangle = |\vec{p}, p^2 = s\rangle \equiv |\vec{p}_s\rangle$ in the plane wave limit. This equation is "half-shell" in the sense that the momentum space wave function solution of equation (1), $\langle \vec{p}|\psi(s)_+\rangle$, has $p^2$ not necessarily equal to $s$. With the definition $T|\phi\rangle = V|\psi_+\rangle$ we can also write the off-shell L–S equation for the amplitude operator $T$ as

$$T(s) = V + VG_0(s)T(s) = V + T(s)G_0(s)V \tag{2}$$

with "off-shell" matrix elements $\langle \vec{p}'|T(s)|\vec{p}\rangle$ found from

$$\langle \vec{p}'|T(s)|\vec{p}\rangle = \langle \vec{p}'|V|\vec{p}\rangle + \int d^3q \langle \vec{p}'|V|\vec{q}\rangle G_0(q;s)\langle \vec{q}|T(s)|\vec{p}\rangle \tag{3}$$

---

†Department of Physics.

(of course, we could also carry out the integration of the (3) in coordinate space). The physical transition probability amplitude is found by putting both $p^2$ and $p'^2$ on-shell as $\langle \vec{p}'_s|T(s)|\vec{p}_s \rangle$. Now of course, equation (3) can also be thought of as half-shell because only the *bra* state of $\langle \vec{q}|$ of $\langle \vec{q}|T(s)|\vec{p} \rangle$ need wander off shell—we shall return to this point later.

We can say that if we take $s$ off the positive real axis and independent of the momenta variables $p'$ and $p$, then equations (1) and (2) are linear integral equations of the second kind.[1] Our goal is to solve such equations systematically with $s$ taking any complex value including above the positive real axis $s \to k^2 + i\varepsilon$. The problems that we must then face are of the following three types:

1.    The positive real axis. As $s \to k^2 + i\varepsilon$, the matrix elements of the "scattering operator" $\langle \vec{p}|K(s)|\vec{p} \rangle$ where $K(s) \equiv G_0(s)V$ become infinite when $p^2 = s$ and the L–S equation becomes a singular integral equation.

2.    The high momentum or small distance behavior of $V$. This may also cause the L–S equation to be singular.

3.    The bound states. As $s \to -|s_B|$ the solutions of the L–S equation also break down.


## 2. OFF-SHELL TECHNIQUES

### Operators and Hilbert Space

We concentrate on the off-shell operator equation (2) which has the formal solutions

$$T(s) = V \frac{1}{I - K(s)} \tag{4}$$

and investigate when such resolvents as $(I - K(s))^{-1}$ are "nonsingular" or "singular."

A list of the formal "functional analysis" properties of operators $K$ and their resolvents $(I - K)^{-1}$ on a linear space of states with norm $\|\psi\|$ includes[2,3]:

1.    *K bounded.* $\|K\| \equiv \mathrm{lub}\,(\|K|\psi\rangle\|/\|\psi\|) < \infty$

2.    *K completely continuous or compact.* $K$ |bounded sequence$\rangle$ $\to$ |compact sequence$\rangle$, which makes $K$ a generalization of a finite matrix in that it has only a point spectrum. Also in some spaces $K$

can be uniformly approximated by dyads.

3. *K Fredholm, Hilbert Schmidt, or $L^2$.* If, $\|K\|_2^2 \equiv \operatorname{Tr} K^+ K \rightarrow$ $\int dp\,dq|K(p, q)|^2 < \infty$ in Hilbert space with norms

$$\|\psi\|_2^2 = \int dp|\psi(p)|^2 < \infty$$

then, $KL^2 \Rightarrow K$ compact $\Rightarrow K$ bounded.

4. Bounded operators imply very little about their resolvents—only if $\|K\| > 1$, then $\|1/(I - K)\| < 1/(1 - \|K\|) < \infty$ and $(I - K)^{-1}$ then exists and is bounded.

5. $K$, compact $\Rightarrow$ Fredholm alternative; either $(I - K)^{-1}$ exists and is bounded or $K|\psi\rangle = |\psi\rangle$ has a nontrivial solution.

6. $K, L^2 \Rightarrow$ the Fredholm form $1/(I - K) = N/D$ where $N$ and $D$ are convergent series and the spectrum occurs when $D = 0$.

For the scattering operator $K(s) = G_0(s)V$ we have the $L^2$ norm,[4]

$$\operatorname{Tr} K^+(s)K(s) \equiv \tau(s) \sim \frac{1}{\operatorname{Im}\sqrt{s}} \int d^3r|V(\vec{r})|^2 \tag{5}$$

which unfortunately is infinite in the scattering region. In fact, $K(s)$ is even unbounded in Hilbert space as $s \rightarrow k^2 + i\varepsilon, \varepsilon \rightarrow 0$, and consequently equation (4) is not acceptable as the solution of the L–S equation if Hilbert space norms are to be used. Moreover, if $s$ is complex and equation (4) is finite, the state norms $\||\vec{p}\rangle\|$ and $\|\langle\vec{p}'|V\|$ are still not in Hilbert space, so the matrix element $\langle\vec{p}'|V\{1/(I-K(s))\}|\vec{p}\rangle$ still does not make any sense.

## $V^{1/2}$ Kernels

If we want a theory which has $L^2$ operators and $L^2$ states in a Hilbert space, we must make a similarity transformation on the L–S equation. Since $G_0(s)$ is diagonal in momentum space and $V$ is diagonal in coordinate space, their square roots make sense and we can choose, (a) $K_G \equiv G_0^{-1/2}KG_0^{1/2}$ which gives an $L^2$ norm[5] of $\tau_G(s) = \operatorname{Tr}\{|G_0(s)|V|G_0(s)|V\}$ which is still infinite as $s \rightarrow k^2 + i\varepsilon$. (b) $\tilde{K}(s) \equiv V^{1/2}K(s)V^{-1/2}$ which gives an $L^2$ norm[6] of $\tilde{\tau}(s) = \operatorname{Tr}\{G_0(s)|V|G_0(s^*)|V|\}$ which does remain finite as $s \rightarrow k^2 + i\varepsilon$. Moreover, evaluation in coordinate space yelds[6,7]

$$\tilde{\tau}(s) = (4\pi)^{-2} \int d^3r\,d^3r' \frac{|V(\vec{r})V(\vec{r}')|}{|\vec{r} - \vec{r}'|^2} e^{-2\operatorname{Im}\sqrt{s'}(\vec{r}-\vec{r}')} \tag{6}$$

$$\bar{\tau}(s) \leqslant \bar{\tau} = (4\pi)^{-2} \int d^3r \, d^3r' \frac{|V(\vec{r})V(\vec{r}')|}{|\vec{r} - \vec{r}'|^2} \tag{7}$$

and if equation (7) remains finite, the $L^2$ norm on the real axis exists, is independent of $s$, and bounds all values of $\bar{\tau}(s)$ for complex $s$. Choice (b) means $\tilde{K}(s) = V^{1/2} G_0(s) V^{1/2}$ so that the dangerous singular Green's function is always protected by two $V^{1/2}$ operators. In this case the solution of the L–S equation (4) is replaced by

$$T(s) = V^{1/2} \frac{1}{I - \tilde{K}(s)} V^{1/2} \tag{8}$$

This solution also handles the problem of states; the matrix elements of $T(s)$ are

$$\langle \vec{p}'|T(s)|\vec{p}\rangle = \left\langle \vec{p}'\left|V^{1/2} \frac{1}{I - \tilde{K}(s)} V^{1/2}\right|\vec{p}\right\rangle \tag{9}$$

so that

$$|\langle \vec{p}'|T(s)|\vec{p}\rangle| \leqslant \|V^{1/2}|\vec{p}'\rangle\| \|V^{1/2}|\vec{p}\rangle\| \left\|\frac{1}{I - \tilde{K}(s)}\right\| \tag{10}$$

and so if $\bar{\tau} < \infty$, then $\|1/(I - \tilde{K}(s))\|$ is finite (except at discrete points) and if we also require

$$\|V^{1/2}|\vec{p}\rangle\| \equiv \bar{c} = (2\pi)^{-3} \int d^3r |V(r)| < \infty \tag{11}$$

then the matrix elements of $T(s)$ are also finite. The states that we must consider in Hilbert space are then weighted by $V^{1/2}$, as $V^{1/2}|\vec{p}\rangle$, $V^{1/2}|\vec{p}'\rangle$, etc.

## Spectral Analysis

Now we turn our attention to the spectrum of $\tilde{K}(s)$, that is, the points at which the resolvent $(I - \tilde{K}(s))^{-1}$ does not exist (discrete only since $\tilde{K}(s)$ is now $L^2$). Since $\tilde{K}(s)$ is also completely continuous, the Fredholm alternative holds; the only poles of $(I - \tilde{K}(s))^{-1}$ [and $\therefore$ of $T(s)$] occur when the eigenvalues of $\tilde{K}(s)$, $\eta_v(s)$ satisfy

$$\eta_v(s) = 1 \tag{12}$$

where

$$\tilde{K}(s)|\psi_v(s)\rangle = \eta_v(s)|\psi_v(s)\rangle \tag{13}$$

since

$$\tilde{K}^\dagger(s) = \tilde{K}(s^*) \tag{14}$$

If the sign of the potential changes a denumerable number of times, (14) does not hold but the conclusions drawn from it remain essentially unchanged.[8] We have the reflection principle

$$\eta_\nu^*(s^*) = \eta_\nu(s) \tag{15}$$

which means equation (12) can only be satisfied when

$$\eta_\nu(-B) = 1 \qquad s = -B < 0 \tag{16}$$

Then equation (13) can be written as

$$-(B + H_0)|\psi_\nu(-B)\rangle = V|\psi_\nu(-B)\rangle \tag{17}$$

the Schrödinger equation for bound states, with $|\tilde{\psi}_\nu(-B)\rangle = V^{1/2}|\psi_\nu(-B)\rangle$.

For general $s$, the eigenstates $|\tilde{\psi}_\nu(s)\rangle$ are related to the eigenstates $|\psi_\nu(s)\rangle$ of the usual Schrödinger equation with $V \to V/\eta_\nu(s)$ at complex energies, where still

$$|\tilde{\psi}_\nu(s)\rangle = V^{1/2}|\psi_\nu(s)\rangle \tag{18}$$

To keep $|\tilde{\psi}_\nu(s)\rangle$ (but not necessarily $|\psi_\nu(s)\rangle$) in Hilbert space we need only have

$$\|\tilde{\psi}_\nu(s)\|^2 = \int d^3r |V(r)| |\langle \vec{r}|\psi_\nu(s)\rangle|^2 < \infty \tag{19}$$

and this weight factor $|V(r)|$ is just what is needed since for large $r$,

$$\langle \vec{r}|\psi_\nu(-B)\rangle \sim \frac{e^{-\sqrt{B}\,r}}{r}$$

becomes for complex $s$

$$\langle \vec{r}|\psi_\nu(s)\rangle \sim \frac{e^{i\sqrt{s}\,r}}{r} \tag{20}$$

Now as $s \to k^2 + i\varepsilon$, equation (20) will not be in Hilbert space, but (19) will be.

Assuming that the eigenfunctions $|\tilde{\psi}_\nu(s)\rangle$ form a complete set,[9] along with the biorthgonality condition

$$\langle \tilde{\psi}_\mu(s^*)|\tilde{\psi}_\nu(s)\rangle = \langle \psi_\mu(s^*)|V|\psi_\nu(s)\rangle \sim \delta_{\mu\nu} \tag{21}$$

we have

$$I = \sum_\nu V^{1/2}|\psi_\nu(s^*)\rangle\langle\psi_\nu(s)|V^{1/2}(\langle\psi_\nu(s^*)|V|\psi_\nu(s)\rangle)^{-1} \tag{22}$$

Contrast this "coupling constant spectrum" with the usual "energy spectrum" which is the spectrum of the non-compact (but Hermitian definite) Hamiltonian operator $H = H_0 + V$ having the completeness relation

$$I = \sum_i |\psi(-B_i)\rangle\langle\psi(-B_i)| + \int_0^\infty ds|\psi(s)_+\rangle\langle\psi(s)_+| \qquad (23)$$

where $|\psi(s)_+\rangle$ is normalized to $\delta(s' - s)$, satisfying equation (1). It is apparent from (22) and (23) that the use of the coupling constant spectrum is a more natural approach to the combined solution of the bound state and scattering problems in that the latter is an analytic continuation of the former; whereas, the use of the energy spectrum mixes the bound state and scattering problems together.

### Resolvent Expansions

Now we expand equation (8) in terms of various convergent series. Since $\tilde{K}V^{1/2} = V^{1/2}K$ the resulting resolvent expansions will look as if the $V^{1/2}$ transformation were never introduced.

1. *Born Series:*  Letting $V \to \lambda V$ and choosing the potential strength so as $|\lambda| \le \|\tilde{K}\|_2^{-1}$ we have

$$T(s) = V^{1/2}(I + V^{1/2}G_0(s)V^{1/2} + \cdots)V^{1/2}$$
$$= V + VG_0(s)V + \cdots = \sum V(G_0(s)V)^n \qquad (24)$$

For the Yukawa potential $V \sim \lambda e^{-r}/r$ and $\tilde{\tau} = |\lambda|^2/2 \Rightarrow |\lambda| \le \sqrt{2}$ and then the Born series always converges.[6]

2. *Fredholm Series:*  $1/(I - \tilde{K}) = I + (\tilde{N}/\tilde{D})$ with $N = \sum \tilde{N}_n$ and $\tilde{D} = \sum \tilde{D}_n = D$

$$\tilde{D}_n = -\frac{1}{n}\, \mathrm{Tr}\,\{\tilde{N}_{n-1} - \tilde{D}_{n-1}\tilde{K}\} = -\frac{1}{n}\sum_{n'=v}^n \tilde{D}_{n-n'}\tilde{K}^n$$

$$\tilde{N}_n = (\tilde{D}_n + \tilde{N}_{n-1})\tilde{K} = \sum_{n'=0}^n \tilde{D}_{n-n'}\tilde{K}^{n'+1}$$

Hence

$$T(s) = V + \frac{V^{1/2}\tilde{N}(s)V^{1/2}}{\tilde{D}(s)} = V + V\frac{N(s)}{D(s)} \qquad (25)$$

convergences for any strength of the potential except when $D(-B_v)$ is zero

3. *Eigenfunction expansion:*  From (4) and (22)

$$T(s) = V + \sum_v \frac{V|\psi_v(s^*)\rangle\langle\psi_v(s)|V}{\langle\psi_v(s^*)|V|\psi_v(s)\rangle}\frac{\eta_v(s)}{1 - \eta_v(s)} \qquad (26)$$

where this expansion may break down near the confluence of two eigenvalues. Note that all three expansions break down at bound state energies $\eta_\nu(-B) = 1$. Moreover, the Born series converges at most only out to $|\eta_\nu(s)| < 1$ for any complex $s$.[4] This gives rise to the study of $\eta$-trajectories and their behavior inside the unit circle $|\eta_\nu(s)| = 1$, and to the interpretation of a resonance as an $\eta_\nu(s)$ lying close to the unit circle, $\eta_\nu(s_{\text{Res}}) \cong 1 + i\delta$. Note also that the matrix elements of these expansions converge uniformly in $\vec{p}$, $\vec{p}'$ and in $s$, which follows in a manner similar to equation (10) because $\tilde\tau$ and $\tilde c$ are independent of $p$, $p'$, and $s$.[9]

4. *Quasiparticle Expansion:* Weinberg[4,6] has given a physical meaning to the separation

$$K_\varrho(s) = \tilde{K}(s) - \sum_a V^{1/2}|a\rangle\langle\tilde{a}|V^{1/2} \tag{27}$$

as the scattering kernel which arises in a theory where all bound states and resonances (near which the Born series diverges) are considered as elementary particles (in $H_0$). Mathematically this is possible because, $\tilde{K}$ is $L^2$ and, therefore, uniformly approximated by dyads. Then

$$T(s) = T_\varrho(s) + \sum_{a,\tilde{a}} T_\varrho(s)|a\rangle\Delta_{a,a}(s)\langle\tilde{a}|T_\varrho(s) \tag{28}$$

where

$$T_\varrho(s) = V^{1/2}\frac{1}{I - K_\varrho(s)}V^{1/2} \tag{29}$$

and the "propagator" is

$$(\Delta^{-1}(s))_{a\tilde{a}} = \delta_{a\tilde{a}} - \langle\tilde{a}|T_\varrho(s)|a\rangle \tag{30}$$

The idea is to choose the dyads $|a\rangle$ and $\langle\tilde{a}|$ such that the eigenvalues are brought within the unit circle so that perturbation theory (the Born series) is again valid.

## Partial Waves

For spherically symmetric potentials $V(r) = V(|\vec{r}|)$, the conditions keeping the Hilbert space norms $\tilde\tau$ and $\tilde c$ finite become

$$\tilde\tau: \int_0^\infty dr\, r|V(r)| < \infty \tag{31}$$

$$\tilde c: \int_0^\infty dr\, r^2|V(r)| < \infty \tag{32}$$

and mean that $V(r) \to \mathcal{O}(r^{-2+\epsilon})$ as $r \to 0$, $V(r) \to \mathcal{O}(r^{-3-\epsilon})$, as $r \to \infty$, $\epsilon > 0$. These same conditions make it possible for us to repeat the entire analysis for each partial wave. Using $T_l = V^{1/2}(I - \tilde{K}_l)^{-1}V^{1/2}$ one can show that the Born, Fredholm, Eigenfunction, and Quasiborn expansions exist and converge uniformly in $p', p', s$, and $l$.[9] Furthermore, make a quasi Born-like subtraction in each partial wave as

$$K_V(s) = \tilde{K}(s) - K_D(s) \tag{33}$$

where the dyadic kernel is $K_D(s) = V^{1/2}G_D(s)V^{1/2}$ with

$$\langle r'l|G_D(s)|rl\rangle = -ikr'rj_l(kr')h_l^{(1)}(kr) \tag{34}$$

$K_V(s)$ is then a Voterra kernel $[\sim\theta(r'-r)]$, which means that $TrK_V^n = 0$, and the Fredholm determinant $\tilde{D}_l(s) = D_l(s)$ is nothing but the Jost function $f_l(k)$.[4]

## The Variational Principle

The condition $\tilde{K}^+(s) = \tilde{K}(s^*)$ enables us to set up a variational principle which is a generalization of the usual Hermitian one,[10]

$$\eta[\psi] = \frac{\langle\tilde{\psi}(s^*)|\tilde{K}(s)|\tilde{\psi}(s)\rangle}{\langle\tilde{\psi}(s^*)|\tilde{\psi}(s)\rangle} = \frac{\langle\psi(s^*)|VG_0(s)V|\psi(s)\rangle}{\langle\psi(s^*)|V|\psi(s)\rangle} \tag{35}$$

or, for example, on partial waves

$$\eta_l[\psi_l] = \frac{\langle\psi_l(s^*)|V|\psi_l(s)\rangle}{\langle\psi_l(s^*)|G_0^{-1}(s)|\psi_l(s)\rangle} \tag{36}$$

where the parameters in a trial wave function as well as the eigenvalue itself are determined by

$$\delta\eta[\psi] = \delta\eta_l[\psi_l] = 0 \tag{37}$$

For the Yukawa potential, a trial wave function of the form[10]

$$\langle p|\psi_l(s)\rangle = \frac{p^{l+1}}{(p^2 - s)\{p^2 + [\mu - i\sqrt{s}\,]^2\}^{l+1}} \tag{38}$$

is an excellent approximation to the lowest eigenstate. It has the correct behavior as $p \to \infty$, $p \to 0$, $p^2 \to s$ and approaches the known Coulomb lowest eignestate as the parameter $\mu \to 0$.

## Separable Approximations

A property of a completely continuous scattering operator near its spectral points is clear from the form of the eigenfunction or quasi Born expansion.

Any potential is effectively separable near the poles of the amplitudes and

$$V \rightarrow \frac{V|\psi\rangle\langle\psi|V}{\langle\psi|V|\psi\rangle} \tag{39}$$

Then we may write near a pole

$$T(s) \approx \frac{V|\psi(s)\rangle\langle\psi(s^*)|V}{\langle\psi(s^*)|V|\psi(s)\rangle\{1 - \eta(s)\}} \tag{40}$$

For real $l$ the pole is a bound state with

$$\eta(s) \approx 1 + (s - s_B)\frac{\partial\eta}{\partial s_B} \tag{41}$$

where

$$\frac{\partial\eta}{\partial s_B} = -\frac{\langle\psi|V|\psi\rangle}{\langle\psi|G_0^{-1}|(\psi)\rangle^2}\langle\psi|\psi\rangle \tag{42}$$

by (36) and $\delta\eta(\psi) = 0$. Hence,[11]

$$T(s_B) \approx \frac{V|\psi\rangle\langle\psi|V}{\langle\psi|\psi\rangle\{s - s_B\}} \tag{43}$$

which agrees with the usual energy spectrum pole if we identify the vertex function $|\Gamma\rangle$ as

$$|\Gamma\rangle = \frac{V|\psi\rangle}{[\langle\psi|\psi\rangle]^{1/2}} \tag{44}$$

For complex $l$ we take $\langle\psi(s^*)| \rightarrow \langle\psi_{l^*}(s^*)|$ in (40) and then near the (Regge) poles,

$$\eta_l(s) \approx 1 + (l - \alpha)\frac{\partial\eta}{\partial\alpha} + \cdots \tag{45}$$

so that

$$T(l, s) \rightarrow \frac{V|\psi_l(s)\rangle\langle\psi_{l^*}(s^*)|V}{\langle\psi_{l^*}(s^*)|\psi_l(s)\rangle\{l - \alpha(s)\}}\left(-\frac{d\alpha}{ds}\right) \tag{46}$$

the usual Regge form.

Wright[10] has obtained the Yukawa–Regge trajectories by use of trial functions equation (38) and the variational principle with $\eta_{l=\alpha(s)}(s) = 1$ determining the Regge poles.

## 3. HALF-SHELL TECHNIQUES†

The $V^{1/2}$ trick leads to a complete operator analysis of the Lippmann–Schwinger equation in Hilbert space. The square root operation and the Green's function "$i\mathcal{E}$" nature of relativistic potential makes this method extremely difficult to carry over to other theories, so now we turn to alternative treatments.

### Banach Space

Hunziker[12] considers the L–S equation in the half-shell form (1) and takes the scattering states in the Banach space (no scalar product defined) $\mathscr{C}$ of the continuous and bounded functions. This has the physical meaning of finite probability densities $|\langle \vec{x}|\phi\rangle|^2$ at position $\vec{x}$ rather than the finite total probabilities $\int d^3x|\langle\vec{x}|\phi\rangle|^2$ of Hilbert space, and corresponds to the conditions in actual experiments.

So we take $|\psi(s)^+\rangle$ and $|\phi(s)\rangle$ in $\mathscr{C}$, where

$$\|\phi\|_x \equiv \sup_x |\langle\vec{x}|\phi\rangle| \tag{47}$$

Hence, the scattering kernel $K(s) = G_0(s)V$ is a bounded operator in $\mathscr{C}$ if

$$\|K(s)\|_x = \sup_{\|\phi\|=1} \|\langle x|K(s)|\phi\rangle\| \leqslant \int d^3y|\langle\vec{x}|G_0(s)|\vec{y}\rangle V(\vec{y})| \tag{48}$$

is finite, which is guaranteed if

$$\int d^3y \frac{|V(\vec{y})|}{|\vec{x} - \vec{y}|} < \infty \tag{49}$$

For spherically symmetric potentials, condition (49) is essentially (31), $\int dr\, rV < \infty$.

Hunziker shows that equation (49) is also enough to guarantee that $K(s)$ is also compact, even on the positive real $s$ axis. This is enough to insure the existence and boundedness of the resolvent $(I - K(s))^{-1}$ of equation (5) but says nothing concerning the Fredholm expansions.

---

†The half-shell techniques have been successfully applied to the scalar Bethe–Salpeter equation and numerical solutions have been obtained Ref: M. Levine, J. Tyon and J. Wright, *Phys. Rev. Letters.* **16**: 962 (1966), and *Phys. Rev.*, to be published.

Since it is the matrix elements of $T(s)$ which physically must remain finite (except at bound states) this analysis requires

$$|\langle\phi|T|\phi\rangle| = |\langle\phi|V|\psi\rangle| \leqslant \|\phi\|_x\|\psi\|_y \int d^3x\, d^3y |\langle\vec{x}|V|y\rangle| \qquad (50)$$

to be finite, and follows if

$$\int d^3r |V(\vec{r})| < \infty \qquad (51)$$

which is condition (12) or (32) $\int dr\, r^2 V < \infty$. So the restrictions on the potential in this analysis are the same as in the $V^{1/2}$ Hilbert space approach.

## Fredholm Reduction

Kowalski and Feldman[13] and Noyes[14] consider the half-shell form of the $T$-matrix L–S equation. Define the partial wave amplitude $\langle \vec{p}|T_l(s)|\vec{p}(s)\rangle \equiv T_l(p, k)$, $k = \sqrt{s}$ so that the L–S equation is an integral equation in the first argument alone,

$$T_l(p, k) = V_l(p, k) + \int dq\, V_l(p, q)G_k(q)T_l(q, k) \qquad (52)$$

We solve the real-axis problem by finding a reduced potential $\bar{V}_l(p, q; k)$ so that $\bar{V}_l(p, q; k) \to 0$ as $q \to k$ thereby removing the singular behavior of the Green's function in the reduced scattering operator $\bar{K} = G\bar{V}$. So we take

$$\bar{V}_l(p, q; k) = V_l(p, q) - \frac{V_l(p, k)V_l(k, q)}{V_l(k, k)} \qquad (53)$$

It is then easy to show that (52) has a solution of the form

$$T_l(p, k) = f_l(p, k)T_l(s) \qquad (54)$$

where $T_l(s) \equiv T_l(k, k)$ is the on-shell amplitude and $f_l(p, k)$ is an off-shell modification form factor, $f_l(k, k) = 1$, satisfying the Fredholm ($L^2$) integral equation

$$f_l(p, k) = \frac{V_l(p, k)}{V_l(k, k)} + \int dq\, \bar{V}_l(p, q; k)G_k(q)f_l(q, k) \qquad (55)$$

Solving this equation systematically (as in Section 2) for $f_l(p, k)$ we find the on-shell amplitude from

$$T_l(s) = \frac{V_l(s)}{1 - \int dq\, V_l(k, q)G_k(q)f_l(q, k)} \qquad (56)$$

Note the interesting features of this procedure:

1.    Equation (53) means we have physically factored the scattering problem into its asymptotic ($r \to \infty$, or $p^2 \to k^2$) part $T_l(s)$ and a part $f_l(p, k)$ depending only on the short-range effects of the interaction.

2.    Equation (54) avoids the real-axis problem but does not improve the $p \to \infty$ (or $r \to 0$) behavior when $V \to \bar{V}$ in equation (52); so the resolvent expansions of equation (54) are guaranteed only if $\int dr \, rV, \int dr \, r^2 V < \infty$. However, Noyes shows that in coordinate space, the $f_l(p, k)$ integral equation has a natural cutoff as $r \to \infty$.

3.    Since $1/T_l(s)$ = principle value integral $- ik$, the amplitude is automatically unitary in any approximation of equation (55). For example, in lowest order

$$T_l(s) \approx \frac{V_l(s)}{1 - (1/V_l(s))\langle k|V_l G_0 V_l|k\rangle} \tag{56}$$

is unitary; whereas, if $V$ itself is weak, equation (55) can be approximated as

$$T_l(s) \approx V_l(s) + \langle k|V_l G_0 V_l|k\rangle \tag{57}$$

the second Born approximation, which is not unitary.

4.    The numerator of equation (54) is the Born approximation and, hence, $T_l(s)$ has the correct threshold $s \to 0$ and asymptotic $s \to \infty$ behavior. Recall that the $N/D$ method can never simultaneously yield the correct threshold and asymptotic behavior.

5.    The poles of (55) occur when $\langle k|V_l G f_l|k\rangle = 1 = \langle k|f|k\rangle$ or at

$$VG_0|f\rangle = |f\rangle \tag{58}$$

which is the bound state Schrödinger equation with vertex function $|f\rangle$. So this formulation does not introduce any spurious poles—at least in the two-body amplitude.

## 4. ON-SHELL ANALYTICITY

Having outlined the various functional analysis techniques which can be applied to potential theory, we now use them to prove the analytic and asymptotic properties of the on-shell amplitudes $\langle \vec{p}_s'|T(s)|\vec{p}_s\rangle \equiv T(s, t)$ where $t \equiv -(\vec{p}_s' - \vec{p}_s)^2$ and $\langle kl|T(s)|kl\rangle = T_l(s)$.

The results, of course, are not new and the reason for doing this here is that the $V^{1/2}$ method alone essentially yields all previous results—easily and with minimal restrictions. We need no longer concern ourselves with questions of uniform convergent resolvent expansions, but instead base the proofs on the following theorems:

*Theorem I:* The operator $K(\xi)$ (or state $|\psi(\xi)\rangle$) is "operator analytic" in $\xi$ if it is uniformly bounded and if its matrix elements (even of continuous variables $|r\rangle$, $|\vec{p}\rangle$, etc.) are analytic in $\xi$ (in the same domain).

*Theorem II:* If $K(\xi)$ is completely continuous and operator analytic in $\xi$, then the resolvent $(I - K(\xi))^{-1}$ is operator analytic in $\xi$ (when it exists).

*Theorem III:* If $K(\xi)$ has a uniformly bounded $L^2$ norm and is operator analytic in $\xi$, then $(I - K(\xi))^{-1} = N(\xi)/D(\xi)$ where $D(\xi)$ is analytic in $\xi$, and $N(\xi)$ is operator analytic in $\xi$.

### Energy Analyticity

Now the states $V^{1/2}|\vec{p}_s\rangle$ are no longer uniformly bounded by $\bar{c}$, equation (11), as $s$ becomes complex and Im $\sqrt{s} \to \infty$, and so we must restrict the domain of analyticity to a region related to the range of ·this potential. For potentials of range $1/m$ define.[7]

$$\bar{c}_m \equiv (2\pi)^{-3} \int d^3r\, e^{(m-\epsilon)r}|V(\vec{r})| \qquad (59)$$

$$\bar{\tau}_m \equiv (4\pi)^{-2} \int d^3r\, d^3r'\, e^{(m-\epsilon)(r+r')} \frac{|V(\vec{r}')V(\vec{r})|}{|\vec{r}' - \vec{r}|^2} \qquad (60)$$

If we restrict ourselves to potentials where $\bar{c}_m$ ond $\bar{\tau}_m$ remain finite then we have

$$\bar{\tau}(s) \leqslant \bar{\tau} < \bar{\tau}_m \qquad (61)$$

for $s$ anywhere on the physical sheet, Im $\sqrt{s} \geqslant 0$, but

$$\|V^{1/2}|\vec{p}_s\rangle\|^2 < \bar{c}_m \qquad (62)$$

only within a very small region of the $s$-plane, $|\mathrm{Re}\,\sqrt{t/4}|^2 + |\mathrm{Im}\,\sqrt{s + t/4}|^2 < m^2/4$.

We can enlarge this region by shifting part of the exponentials in

$$T(s, t) = (2\pi)^{-3} \int d^3r\, d^3r'\, e^{-i\vec{p}_s\cdot\vec{r}'} e^{i\vec{p}_s\cdot\vec{r}} \langle\vec{r}'|V^{1/2}(I - (\tilde{K}(s))^{-1} V^{1/2}|\vec{r}\rangle \qquad (63)$$

into the operators themselves

$$T(s, t) - V(t) = \langle \varphi'(s, t)|K_1(s, t)[I - K_2(s, t)]|\varphi(s, t)\rangle \quad (64)$$

until the analogous restrictions to (61) and (62) applied to $K_1(s, t)$, $K_2(s, t)$, $|\varphi(s, t)\rangle$ and $|\varphi'(s, t)\rangle$ yield only one restriction. This we find is the Hunziker domain

$$\left|\text{Re} \frac{\sqrt{t}}{4}\right|^2 + \left\{\text{Im} \frac{\sqrt{s + t}}{4} - \lambda \text{ Im} \sqrt{s}\right\}^2 < m^2 \qquad -1 \leqslant \lambda \leqslant 1 \tag{65}$$

obtained by the Banach space approach.[12] Finally, because of Theorems I, II, and III, equation (65) implies that $T(s, t)$ is analytic on the entire $s$ physical sheet (except for poles) when $t$ is restricted within the circle

$$|t| < 4m^2 \tag{66}$$

The original proofs[15,16] contained complications due to uniform convergence and were only valid on the line $0 \leqslant t \leqslant -4m^2$.

### Energy Asymptotics

The proof of Klein and Zemach that

$$|T(s, t) - V(t)| \to 0 \tag{67}$$

as $|s| \to \infty$ was valid only on the line $0 \leqslant t \leqslant -4m^2$. We can extend this to the circle $|t| < 4m^2$ by using $V^{1/2}$ norms as in the last section. This is accomplished by expanding $T(s, t)$ into the Born series and then integrating by parts to pick up a convergence factor $1/\sqrt{s}$, as $|s| \to \infty$.[16]

### Momentum Transfer Analyticity

Clearly, from (64), $T(s, t) - V(t)$ is also analytic in $t$ for $|t| < 4m^2$ for any $s$, and for $t$ within the large Lehmann Ellipse cosh $(\text{Im} \theta_s) < 1 + 4m^2/2s$ in the cos $\theta_s = \hat{p}'_s \cdot \hat{p}_s$ plane for $s$ real.

To enlarge this domain of analyticity to the entire finite plane (except for the cut from $t = 4m^2$ to $\infty$), we must, of course, specialize $V(t)$ to the usual sum of Yukawa potentials,

$$V(t) \sim \int_{m^2}^{\infty} dt' \frac{\sigma(t')}{t' - t} \tag{68}$$

The proof then follows according to Blankenbecler *et al.*[17] Use the energy shell Fredholm expansion

$$T(s, t) - V(t) = \frac{1}{D(s)} \sum_{n=0}^{\infty} \sum_{n'=0}^{\infty} D_{n-n'}(s) B_{n'+1}(s, t) \tag{69}$$

where $B_n(s, t)$ is the $n$th Born term $\langle \vec{p}_s' | V(G_0(s)V)^n | \vec{p}_s \rangle$ and transmit the $t$ analyticity of $B_n(s, t)$ to $T(s, t) - V(t)$ which follows if the Fredholm series converges uniformly for $t$ in the physical region. This assumption is easily proved using $V^{1/2}$ norms as was noted in Section II.

## Partial wave Analyticity Asymptotics

The analytic structure of $T_l(s)$ (for Yukawa potentials) can be obtained by a similar analysis on the $V^{1/2}$ norms, $\bar{\tau}_l(s)$ and $\bar{c}_l(s)$. However, a more thrifty approach follows from the Jost form

$$T_l(s) \sim S_l(s) - 1 \tag{70}$$

with[5,18]

$$S_l(s) = \frac{D_l(\bar{s})}{D_l(s)} e^{-i\sqrt{s} V_l(s)} \tag{71}$$

and

$$V_l(s) = Q_l\left(1 + \frac{m^2}{2s}\right) \tag{72}$$

and $D_l(\bar{s})$ is the Fredholm determinant (or Jost function) on the second sheet. For real $l$ the $V^{1/2}L^2$ norm is, for $l > -3/2$, $l \neq -1$,

$$\bar{\tau}_l(s) = \int_0^{\infty} dp\, dq\, \frac{\left|Q_l\left(\frac{p^2 + q^2 + m^2}{2pq}\right)\right|^2}{(p^2 - s)(q^2 - s^*)} < \infty \tag{73}$$

which remains finite as $s \to s_R + i\varepsilon$ and vanishes as

$$\bar{\tau}_l(s) \to O(|s|^{-1}) \text{ as } |s| \to \infty \tag{74}$$

For complex $l$, $\bar{\tau}_l(s)$ given by (73) is not quite the correct $L^2$ norm[19] but nevertheless it does bound the actual terms occurring in the Fredholm expansions, $\operatorname{Tr} K^n$, as was noted recently by Wright.[20] Then,

$$\bar{\tau}_l(s) \to O(|l|^{-1}) \quad \text{as } |l| \to \infty \tag{75}$$

for $\operatorname{Re} l \geqslant -1$, arbitrary $s \neq 0$, and $\operatorname{Re} l > -1/2$ for $s = 0$.

We can immediately conclude the usual analyticity structure of $T_l(s)$ in both the $s$ and $l$ planes, as well as the asymptotic behavior

$$\left|\frac{T_l(s)}{V_l(s)} - 1\right| \to 0 \tag{76}$$

since $D_l(s) \to 1 + O(\bar{\tau}_l(s))$, as either $|s| \to \infty$ or as $|l| \to \infty$, $\mathrm{Re}\, l \geqslant -1/2$, and this includes the very tricky region $\mathrm{Im}\, l \to \infty$, which prevented Regge $et$ $al.$[21] from completing their proof of the Mandelstam representation starting with the partial wave analytic structure due to the Regge poles at $D_{l=\alpha}(s) = 0$.

## ACKNOWLEDGMENTS

The author would like to thank Dr. S. Weinberg and Dr. J. Wright, as much of this work has been done in conjunction with them.

## REFERENCES

1. F. Smithies, "Integral Equations," Cambridge University Press, New York, 1958.
2. F. Reisz and B. Sz-Nagy, "Functional Analysis," Ungar, New York, 1955.
3. S. Weinberg, *Phys Rev.* **133**: B232 (1964).
4. S. Weinberg, *Phys. Rev.* **131**: 440 (1963).
5. L. Brown, D. Fivel, B. Lee, and R. Sawyer, *Ann. Phys.* **23**: 187 (1963).
6. M. Scadron, S. Weinberg, and J. Wright, *Phys. Rev.* **135**: B202 (1964).
7. A. Grossman and T. Wu, *J. Math. Phys.* **2**: 710 (1961).
8. K. Meetz, *J. Math. Phys.* **3**: 690 (1962).
9. M. Scadron, Thesis, UCRL Report 11566.
10. J. Wright and M. Scadron, *Nuovo Cimento* **34**: 1571 (1964).
11. J. Wright, Thesis, UCRL Report. Feb 1965.
12. W. Hunziker, *Helv. Phys. Acta* **34**: 593 (1961).
13. K. Kowalski and D. Feldman, *J. Math. Phys.* **4**: 507 (1963).
14. H.P. Noyes, *Phys. Rev. Letters* **15**: 538 (1965).
15. N. Khuri, *Phys. Rev.* **107**: 303 (1957).
16. A. Klein and C. Zemach, *Ann. Phys.* **1**: 440 (1959).
17. R. Blankenbecler, M. Goldberger, N. Khuri, and S. Treiman, *Ann Phys.* **10**: 62 (1960).
18. M. Scadron and J. Wright, *Nuovo Cimento* **37**: 1747 (1965).
19. S. Kearsfeld and M. Bertero, private communication.
20. J. Wright, private communication.
21. S. Bottino, A. Longoni, and T. Regge, *Nuovo Cimento* **23**: 954 (1963).

# Some Simple Bootstrap Models

B. M. UDGAONKAR

*TATA INSTITUTE OF FUNDAMENTAL RESEARCH*
*Bombay, India*

## 1. INTRODUCTION

The idea of bootstrap as a possible mechanism underlying the dynamics of strong interactions arose in the early work of Chew and Mandelstam[1] on $\pi-\pi$ scattering, over five years ago. Since then, a large number of bootstrap calculations have been made for different kinds of systems by using a variety of approximations.† These have made it plausible that bootstraps may provide the basic mechanism for binding the large number of hadrons we know today, and that approximations based on nearby singularities and a suitable parametrization of short-range effects—cutoffs, Balázs poles etc.—are reasonably adequate for many purposes. We shall discuss a class of simple models based on these ideas. We do not expect these to give quantitative results for masses of particles, which appear to depend on the details of the short range forces. So in these models we shall not mention any predictions about the masses. On the other hand, the information they furnish on multiplet structure, particle spectra, and relative couplings is expected to be more reliable, and we shall concentrate on these aspects.

These models are in a sense a natural generalization of Chew's reciprocal bootstrap[2] for $N$ and $N^*$. We shall, therefore, begin by

---

†For a recent review and references see B. M. Udgaonkar in "Bootstraps," Seminar on High-Energy Physics and Elementary Particles, Trieste, 1965 (IAEA, Vienna).

reviewing the static $N/D$ model with linear $D$ approximation, and see how the condition for this bootstrap can be written in a simple elegant fashion in terms of the relevant crossing matrix.

## 2. RECIPROCAL BOOTSTRAP OF $N$ AND $N^*$

In Chew's reciprocal bootstrap for $N$ and $N^*$ one considers $p$-wave $\pi$-$N$ scattering in the static approximation. Forces are generated by the exchange of the four particle states with $(IJ) = (1/2, 1/2), (1/2, 3/2), (3/2, 1/2)$ and $(3/2, 3/2)$, that can possibly exist (Fig. 1) in the $u$-channel and one finds that the simplest solution is one which involves $N$ and $N^*$ alone.

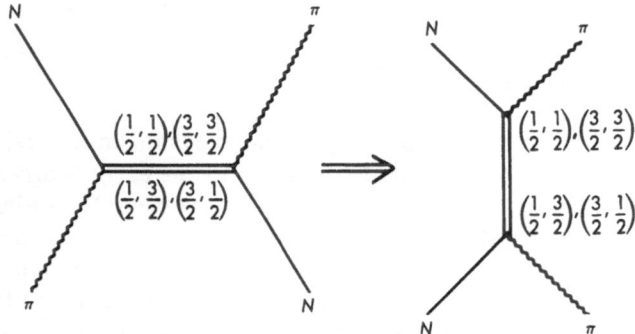

Fig. 1. Dynamics of $p$-wave $\pi$-$N$ scattering.

Let us look at this calculation in some detail. We shall specify a $p$-wave meson–baryon state by its isospin $I$, spin $J$, and energy $W$. We use the amplitude

$$g_{IJ}(\omega) = e^{i\delta_{IJ}} \frac{\sin \delta_{IJ}}{q^3} \tag{1}$$

where $\delta_{IJ}$ is the phase shift in the $(IJ)$-state, $q^2 = \omega^2 - 1$, and $\omega = W - m$, and we take the meson mass as unity. The forces are obtained from the crossing relation

$$g_{IJ}(s) = \sum_{I'J'} \alpha_{II'} \beta_{JJ'} g_{I'J'}(u) \tag{2}$$

where $\alpha$ and $\beta$ are the crossing matrices for isospin and spin, respectively, $I, I'$ and $J, J'$ take the values $1/2$ and $3/2$. The crossing relation (2) is a linear relation which expresses an amplitude of given isospin and spin $(IJ)$ in one channel in terms of amplitudes of definite isospin and spin in the crossed $u$-channel. In the static limit, $s \rightarrow u$ corresponds to $\omega \rightarrow -\omega$, and equation (2) may be rewritten as

$$g_{IJ}(\omega) = \sum_{I'J'} \alpha_{II'} \beta_{JJ'} g_{I'J'}(-\omega) \tag{3}$$

If some isobar (a bound state or resonance) occurs in the $(IJ)$-state, the corresponding amplitude has a pole

$$\frac{\gamma_{IJ}}{\omega_{IJ} - \omega}$$

From equation (3), the Born term which gives the force in the $(IJ)$-state due to the various exchanges in the $u$-channel is

$$B_{IJ}(\omega) = \sum_{I'J'} \alpha_{II'} \beta_{JJ'} \frac{\gamma_{I'J'}}{\omega_{I'J'} + \omega} \tag{4}$$

For the moment, we shall let the sum on the right-hand side go over all the possible values of $I'$ and $J'$, namely, $1/2$ and $3/2$. If there is no particle in any particular state $(I'J')$, the corresponding $\gamma_{I'J'}$ will be zero. We now use this $B_{IJ}(\omega)$ as the input in an $N/D$ calculation to obtain

$$g_{IJ}(\omega) = \frac{N_{IJ}(\omega)}{D_{IJ}(\omega)} \tag{5}$$

$$N_{IJ}(\omega) = \sum_{I'J'} \alpha_{II'} \beta_{JJ'} \frac{\gamma_{I'J'} D_{IJ}(-\omega_{I'J'})}{\omega_{I'J'} + \omega} \tag{6}$$

and

$$D_{IJ}(\omega) = 1 - \frac{\omega - \omega_0}{\pi} \int_1^A d\omega' \frac{(\omega'^2 - 1)^{3/2} N_{IJ}(\omega')}{(\omega' - \omega_0)(\omega' - \omega)} \tag{7}$$

where $\omega_0$ is some substraction point and $\Lambda$ is a cutoff parameter which parametrizes our ignorance of short-range and high-energy effects. These equations ensure the correct elastic unitarity and at the same time give the correct force singularities arising from the $B_{IJ}$ contribution to $g_{IJ}$.

Now suppose we have an isobar in the state $(IJ)$. Then guided by one's experience from a large number of calculations where one has found that the $D$-function is approximately linear in the neighborhood of a bound state or resonance, we introduce the linear $D$ approximation, i.e., we set

$$\text{Re } D_{IJ}(\omega) = \frac{\omega_{IJ} - \omega}{\omega_{IJ} - \omega_0} \tag{8}$$

Then,

$$\gamma_{IJ} = -\frac{N_{IJ}(\omega_{IJ})}{\operatorname{Re} D'_{IJ}(\omega_{IJ})} = \sum_{I'J'} \alpha_{II'}\beta_{JJ'}\gamma_{I'J'} \tag{9}$$

It is to be noted that the $\gamma$'s on the left-hand side of this equation are the "output" $\gamma$'s while those on the right-hand side are the "input" $\gamma$'s representing the strength of the input forces. Thus, equation (9) is the self-consistency condition that output = input. We shall rewrite this condition in the matrix form as

$$\Gamma = C\Gamma \tag{10}$$

where $\Gamma$ is a column vector whose elements are the $\gamma_{IJ}$ and $C$ is the crossing matrix $\alpha \otimes \beta$. Equation (10) will be the basic equation for the models which we shall discuss.

For the $p$-wave $\pi-N$ problem under consideration, one knows the crossing matrices to be

$$\begin{array}{cc} & \begin{array}{cc} 1/2 & 3/2 \end{array} \\ \alpha = \beta = \begin{pmatrix} -1/3 & 4/3 \\ 2/3 & 1/3 \end{pmatrix} & \begin{array}{c} 1/2 \\ 3/2 \end{array} \end{array} \tag{11}$$

With these crossing matrices, one can readily see that equation (10) is equivalent to the following two conditions on the coupling constants:

$$\gamma_{(1/2)(3/2)} = \gamma_{(3/2)(1/2)} \tag{12}$$

$$2\gamma_{(3/2)(3/2)} = \gamma_{(1/2)(1/2)} + \gamma_{(1/2)(3/2)} \tag{13}$$

The simplest solution of these equations, i.e., one involving the smallest number of particles, is one with

$$\gamma_{(1/2)(3/2)} = \gamma_{(3/2)(1/2)} = 0 \tag{14}$$

and

$$2\gamma_{(3/2)(3/2)} = \gamma_{(1/2)(1/2)} \tag{15}$$

i.e., one with particles only in the $(I, J) = (1/2, 1/2)$ and $(3/2, 3/2)$ states (namely, $N$ and $N^*$), and none in the $(I, J) = (1/2, 3/2)$ or $(3/2, 1/2)$ states. The couplings of $N$ and $N^*$ are then related by equation (15) which is known to agree quite well with experiment. Also, the amplitude evaluated in this approximation reproduces the well-known Chew–Low effective range formula, namely,

$$\frac{4}{3} f^2 \frac{q^3}{\omega} \cot \delta_{(3/2)(3/2)} = 1 - \frac{\omega}{\omega_{33}} \tag{16}$$

This success of the static $N/D$ model with linear approximation for the $D$ function in describing the experimental situation in low energy $\pi$-$N$ scattering encourages one to use it for a description of other meson–baryon scattering processes. We shall now proceed to do so.

## 3. THE STRONG COUPLING SERIES

The first extension of the Chew reciprocal bootstrap that we shall consider is the work of Abers, Balázs, and Hara (ABH).[3] These authors considered a succession of scattering problems involving pions and nucleon isobars, and thus suggested the existence of a series of isobars characterized by $I = J = 1/2, 3/2, 5/2, \ldots, \infty$. This series has been well known for a long time from the work of Wentzel and others[4] who had obtained it as an exact solution of the static pion–nucleon scattering problem in the strong coupling limit of $g \to \infty$. It is therefore known as the strong coupling series. ABH, however, did not make the strong coupling assumption in their derivation of the series, at least not explicitly. In reviewing their work, therefore, we shall try to see if something similar to this assumption was actually involved.

The first step in ABH is to consider the $p$-wave scattering of pions by $N^*$, the nucleon isobar of $I = J = 3/2$. The possible states in this scattering are characterized by $I = 1/2, 3/2, 5/2$ and $J = 1/2, 3/2, 5/2$. The states with $I, J = 1/2, 3/2$ will communicate with the lower lying $\pi$-$N$ channel, and will, therefore, have to be included in a two channel calculation. The states with $I$ and/or $J = 5/2$, on the other hand, will appear only in a single channel problem at this stage, though they will also be involved in multichannel problems when higher isobars are taken into account. ABH neglect such channels which will have a higher threshold, and consider the problem of the $I$ and/or $J = 5/2$ states as a single channel problem of $\pi$-$N^*$ scattering. Then it is easy to see from the relevant crossing matrices and the equation (10) that one can predict a particle in the state with $I = J = 5/2$. Calling it $N^{**}$, one can then consider $\pi$-$N^{**}$ scattering, again as a one-channel problem as far as states with $I$ and/or $J = 7/2$ are concerned, and is thus led to predict a particle in the state $I = J = 7/2$, and so on. Thus, considering a series of one-channel problems in which the higher isobars are assumed not to affect the calculation of lower isobars produced in single channel calculations, they arrive at the strong coupling series $I = J = 1/2, 3/2, 5/2, \ldots, \infty$. At the same time, they also derive the ratios of the coupling constants of the isobars in this series.

A question that naturally arises now is whether the treatment of meson isobar scattering as a series of single channel scattering problems in this fashion is justified, and if so under what conditions. This became all the more pertinent when it was found, from the work of Cook, Goebel, and Sakita (CGS),[5] that the strong coupling series could also be derived from the algebraic properties of the multi-channel Chew–Low equation in the limit $g \to \infty$, and further that the coupling constant ratios in the CGS model, as evaluated by Singh,[6] were precisely the same as those derived by ABH.

The relationship between the single channel and multichannel calculations within this kind of static models was clarified by A. Kumar,[7] who showed that the multichannel problem of meson isobar scattering in the static model decouples at the poles into a series of single channel problems if one either assumes that all the isobars are mass-degenerate, or makes some simple assumptions about the $D$-matrix. To be more specific, under these assumptions, the multichannel problem leads to equations of the form $\Gamma = C\Gamma$ for each single channel. The results of a complete multichannel bootstrap are then the same as those obtained by the ABH procedure. Further, since the masses of isobars would be degenerate in the limit of $g \to \infty$, one also gets some idea, in the mass degenerate case, as to why the strong coupling series arises in the work of ABH.

At this stage it is still not clear as to why the results of bootstraps based on $\Gamma = C\Gamma$ coincide with the results obtained from the algebraic considerations of CGS. Let me now first describe briefly the CGS model. CGS start with the Chew–Low equation for the scattering amplitude, which may be written as

$$
\begin{aligned}
T_{\beta\alpha}^{JI}(\omega) = &-\lambda^2 \sum_K \left\{ \frac{(A_\beta)^{JK}(A_\alpha)^{KI}}{M_K - M_I - \omega} + \frac{(A_\alpha)^{JK}(A_\beta)^{KI}}{M_K - M_J + \omega} \right\} \\
&- \sum_{p,K} \left\{ \frac{[T_{\gamma\beta}^{KJ}(\omega_p)]^* T_{\gamma\alpha}^{KI}(\omega_p)}{M_K + \omega_p - M_I - \omega} \right. \\
&\left. + \frac{[T_{\gamma\alpha}^{KJ}(\omega_p)]^* T_{\gamma\beta}^{KI}(\omega_p)}{M_K + \omega_p - M_J + \omega} \right\} + \cdots
\end{aligned}
\tag{17}
$$

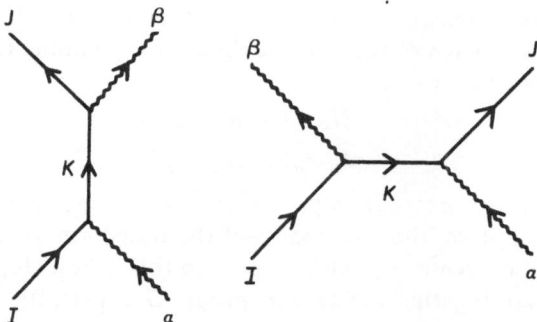

Fig. 2. Pole diagrams contributing to equation (17).

where the first two terms correspond to the pole diagrams of Fig. 2. Here the baryon indices $I, J, K, \ldots$, stand for index pairs $(i, i_z), \ldots$ representing the isospin and its third component, and the meson indices $\alpha, \beta, \ldots$ stand for the third component of the meson isospin. For the sake of economy in writing, we are here considering the s-wave scattering of a scalar isovector meson by baryon isobars. But the results are readily generalized to the p-wave scattering of a pseudoscalar isovector meson, which is the case of physical interest. $\lambda(A_\alpha)^{JI} \equiv \lambda A^{jj_z;ii_z}$ describes the coupling of the meson $\alpha$ to the baryons $(i, i_z)$ and $(j, j_z)$ (Fig. 3).

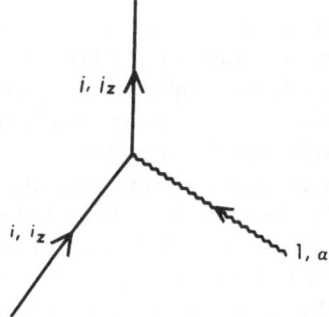

Fig. 3. Diagram defining the meson-baryon coupling $\lambda(A_\alpha)^{JI}$.

The requirement that the amplitude remain finite in the strong coupling limit $(\lambda^2 \to \infty)$ gives the dynamical equation of the CGS theory, namely,

$$[A_\alpha, A_\beta] = 0 \tag{18}$$

Here $A_\alpha$ is a matrix whose $JK$-th element is $(A_\alpha)^{JK}$, i.e., it is a matrix in the space of the isobars. CGS now combine this equation with the two equations

$$[I_i, I_j] = i\epsilon_{ijk}I_K \qquad (19)$$

$$[I_i, A_\alpha] = i\epsilon_{i\alpha\beta}A_\beta \qquad (20)$$

which describe, respectively, the invariance group (namely, the isospin group) of the problem, and the transformation property of the meson currents $A_\alpha$ with respect to this group. Equations (18), (19), and (20) together define the group $G = [SU(2)]_I \times T_3$ which is a semidirect product of the isospin group $[SU(2)]_I$ which is the invariance group of the system under consideration, and a translation group in three dimensions. The isobar spectrum is a representation of this larger group $G$. Since this spectrum generating group $G$ is noncompact, the spectrum of isobars is infinite dimensional.

Coming now to the case of physical interest, that of $p$-wave scattering of pseudoscalar mesons, the meson current operators may be denoted by $A_{i\alpha}$; $i = 1, 2, 3$ for the three orbital angular momentum states corresponding to $l = 1$, and $\alpha = 1, 2, 3$ for the three isospin states. Equation (18) now gets replaced by

$$[A_{i\alpha}, A_{j\beta}] = 0 \qquad (21)$$

and the spectrum generating group becomes

$$G = ([SU(2)]_I \otimes [SU(2)]_J) \times T_9$$

The representations of this group were derived by CGS by the method of group contraction on $SU(4)$, and the simplest of these was found to be one whose members were characterized by $I = J = 1/2, 3/2, 5/2, \ldots, \infty$, i.e., the strong coupling series. Further, as mentioned earlier, the coupling constant ratios of the CGS model, which can be derived readily from the algebra, were found by Singh to be precisely the same as in the ABH model. The latter, as we have seen, coincides with the multichannel bootstrap model based on equation (10), if the masses of isobras are assumed to be degenerate, or if the $D$-matrix is assumed to have some simple properties.

It is not difficult now to understand why the results of the CGS model coincide with those obtained from conventional bootstraps based on equation (10). The crucial equation in the CGS model is the equation (21) representing the strong coupling condition, the other equations [equations (19), (20), or their analogues] being just the equations describing the invariance group of the system. One, therefore, suspects that equation (21) is equivalent to the bootstrap condition of equation (10). This is indeed the case,[8] and one can exploit this equivalence to construct representations of the strong coupling group $G$. This was the method used by Singh and Udgaonkar[8] for deriving the strong coupling series for hyperon isobars. This series is characterized by

$$J = I \pm \frac{1}{2} = \frac{1}{2}, \frac{3}{2}, \frac{5}{2}, \ldots, \infty$$

and cannot be obtained by the method of group contraction† on $SU(4)$. The lowest members of this series may be identified with $\Lambda$, $\Sigma$ and $Y_1^*(1385)$. The hyperon-pion coupling constants predicted by the model are in reasonable agreement with experiment or $SU(3)$ predictions or both. For example, one predicts the branching ratio

$$\frac{\Gamma(Y_1^* \to \Sigma\pi)}{\Gamma(Y_1^* \to \Lambda\pi)} = 12\%$$

to be compared with the experimental value of $11 \pm 2\%$ and the $SU(3)$ prediction of $16\%$. Isovector magnetic moments can also be calculated and one finds,[8] for example, that

$$\left(\frac{\mu_\Sigma}{\mu_{Y_1^*}}\right)_V = \frac{2}{3}$$

in agreement with the prediction of $SU(6)$.

## 4. EXTENSION TO HIGHER SYMMETRIES

The first extension to higher symmetries that we shall consider will be the $SU(3)$ analogue of the $N-N^*$ bootstrap. If we consider the scattering of the pseudoscalar meson octet by the baryon octet, we would expect the baryon octet (which we shall denote by $8_{1/2}$) and the spin-3/2 decimet ($10_{3/2}$) to bootstrap each other. This they do.[9] However, the bootstrap between them is only approximate in the sense of our bootstrap condition (10), unlike the "exact" bootstrap between the $N$ and $N^*$. If one insists on an "exact" bootstrap

---

†It can, however, be obtained by starting with $SU(2) \otimes SU(4)$ as the uncontracted group, or from the inhomogeneous Euclidean group in four dimensions. See P. Babu, A. Rangwala, and V. Singh, *Phys. Rev.* **157**: 1322 (1967).

in this case, one has to include the multiplets $\overline{10}_{1/2}, 27_{1/2}$, and $27_{3/2}$ also, as was shown by Goebel and by Singh.[10] The couplings of the $\overline{10}_{1/2}, 27_{1/2}$, and $27_{3/2}$, however, turn out[11] to be small compared to those of $8_{1/2}$ and $10_{3/2}$, and, hence, one can consider the particles in the states $\overline{10}_{1/2}, 27_{1/2}$, and $27_{3/2}$ to be inessential for the dynamics of the $8_{1/2}$–$10_{3/2}$ pair and treat them as "dependent" particles.†
Moreover, one can show that of all the bootstrap possibilities arising in the case of pseudoscalar meson octet baryon octet scattering, the one involving the $8_{1/2}$ and $10_{3/2}$ is the simplest or minimal.[11]

An extension of this bootstrap to $SU(6)$ was considered by Capps,[12] by Belinfante and Cutkosky[12] and by Singh and Udgaonkar.[13] It was found that in the $p$-wave scattering of a meson 35-plet by a baryon 56-plet, the 56-plet bootstraps itself approximately. Further, the *56* self-bootstrap is the simplest[13] bootstrap possibility for baryons in $SU(6)$. If one wants this bootstrap to be "exact" in the sense of equation (10), one has to include other $SU(6)$ multiplets besides the *56*, in particular the 700-plet.

Abers, Baláazs, and Hara have considered the $SU(3)$ analogue of the strong coupling series. Thus, they show that when one considers meson–$10_{3/2}$ scattering one is led to predict the existence of the multiplet $35_{5/2}$, that meson–$35_{5/2}$ scattering leads to the isobar $81_{7/2}$ and so on. The complete series can be found in an elegant fashion by using the strong coupling model of CGS, which, as we have seen, is equivalent to a bootstrap model based on equation (10). This has been done by Goebel.[14] The first few members of the series are

$$8_{1/2}, \ 10_{3/2}, \ \overline{10}_{1/2}, \ 27_{3/2}, \ 35_{5/2}, \ 27_{1/2}, \ 64_{5/2}, \ 35_{3/2}, \ 81_{7/2}, \ \ldots$$

## 5. CAN ONE HAVE A BOOTSTRAP WITH ONLY A FINITE NUMBER OF BARYON MULTIPLETS?

We have, thus, been led again to an infinite series of isobars. Now while one might say that this kind of dynamical model is a reasonable description of reality only for the low-lying states, or in the limit of $g^2 \to \infty$, it would be nice if one could construct a model in such a way that one got only the low-lying isobars, e.g., only $N$ and $N^*$ in the $SU(2)$ case, or only $8_{1/2}$ and $10_{3/2}$ in the $SU(3)$ case. One suggestion in this direction was that for finite $g$, one should

---

†See footnote 13 in the reference quoted in the footnote on p. 117.

consider an intermediate coupling group, which if compact, would have finite dimensional representations. In the $SU(2)$ case, CGS suggested $SU(4)$ as a possible intermediate coupling group, and this was investigated in detail by Kuriyan and Sudarshan.[15] There is, however, one undesirable feature in $SU(4)$ as an intermediate coupling group: it would allow the nucleon to exist by itself (i.e., without even the $N^*$) if $g$ is sufficiently small, whereas, we would expect the simplest physical system to be one consisting of $N$ and $N^*$, if we believe in the reciprocal bootstrap as the basic dynamics of the $\pi-N$ system.

Another possibility, suggested through the work of Fulco and Wong,[16] is the inclusion of forces arising from exchanges in the $t$-channel, i.e., meson exchange forces, which so far have been completely neglected.

Fulco and Wong, in an attempt to establish a correspondence between the results obtained from static bootstrap dynamics and those derived from higher symmetries or current algebras, consider-ed the scattering of an axial vector meson octet $(A_8)$ by a baryon octet. The exchanges included were those of the baryon multiplets $8_{1/2}$ and $10_{3/2}$ in the $u$-channel, and of the meson multiplets $A_8$, $A_1$ and $V_8$ in the $t$-channel, where $A_1$ is an axial vector meson singlet and $V_8$ is a vector–meson octet (Fig. 4). In order to accommodate

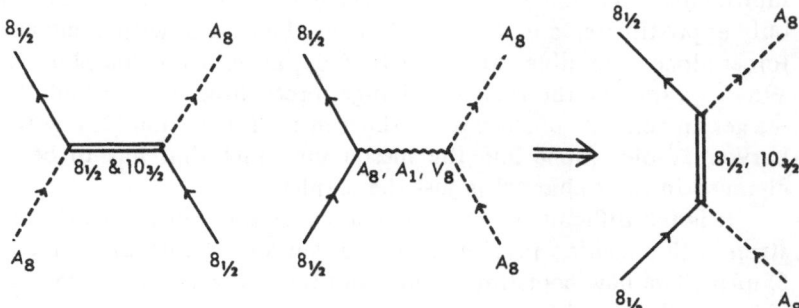

Fig. 4. The Fulco-Wong model.

the meson exchanges, the bootstrap condition (10) was modified to read

$$\Gamma = C_{su}\Gamma + C_{st}\Gamma' \tag{22}$$

where $C_{su}$ is the crossing matrix connecting the $u$ and $s$ channels, and $C_{st}$ is the crossing matrix connecting the $t$ and $s$ channels. $\Gamma'$

is the column vector of coupling constants corresponding to the meson exchange diagram. Inclusion of meson exchange forces in this fashion amounts to approximating to the complicated singularity structure corresponding to meson exchanges by means of simple poles, and is a rather drastic approximation which has no simple justification in terms of a static model as for baryon exchanges. However, granting this, one finds that the new bootstrap condition (22) is satisfied exactly with the baryons $8_{1/2}$ and $10_{3/2}$ only, i.e., without having to include the $\overline{10}_{1/2}$, $27_{1/2}$, $27_{3/2}$, or any other baryon multiplets. Further, one finds that there is no solution involving $8_{1/2}$ as the only bound state, i.e., one not involving the $10_{3/2}$. One must have both $8_{1/2}$ and $10_{3/2}$.

This is very nice as far as the baryons are concerned, but the model looks rather artificial since it is built in terms of the scattering and exchanges of axial vector mesons. However, it does suggest that the $t$-channel forces may not be unimportant and that their inclusion may yield a closed system containing only a small number of baryon multiplets.

Motivated by the Fulco–Wong model, we shall now consider the scattering of a meson 35-plet by a baryon 56-plet in $SU(6)$ with the inclusion of the forces arising from the exchange of a baryon 56-plet and various allowed, but for the moment unspecified, meson multiplets.[17] We know[13] that the self-bootstrap of the 56-plet is only approximate, and that an "exact" bootstrap with $u$-channel forces alone is possible only if one includes at least the 700-plet also. When we include the meson-exchange forces, however, we find that we get an "exact" bootstrap in the sense of equation (22) with a baryon 56-plet alone and the meson multiplet that has to be exchanged in the $t$-channel is just the 35-plet.

It is not difficult to see why one is able to get a closed bootstrap in this fashion involving only a baryon 56-plet and a meson 35-plet. The new bootstrap condition (22) can be shown to be equivalent to the condition

$$[A_\alpha, A_\beta] = C^\gamma_{\alpha\beta} A_\gamma \tag{23}$$

when the external mesons as well as the exchanged mesons belong to a 35-plet. Here the $A_\alpha$'s are the 35 meson currents and $C^\gamma_{\alpha\beta}$ are the structure constants of $SU(6)$. Hence, equation (22) can be satisfied with a baryon multiplet belonging to any representation of $SU(6)$.

At this point we come back to our earlier remark that equation (22) has no simple justification in terms of the conventional static model in the same sense that equation (10) has. What one has done in writing it is to make the plausible ansatz that the left-hand cuts due to the $t$-channel exchanges may be represented by effective poles. Once the $t$-channel exchanges are considered in this fashion, there is no reason why the $u$-channel exchanges also may not be so regarded, and we are thus encouraged to apply equation (22) to more general situations where the static approximation is no longer applicable, in particular to meson–meson scattering. When we do this for the $p$-wave meson–meson scattering in $SU(6)$, we may rewrite equation (22) as

$$\gamma = C_{su}\gamma + C_{st}\gamma \tag{24}$$

where $\gamma$ is now the column vector whose elements are the squares of meson–meson couplings. We then find that equation (24) is satisfied with a meson 35–plet by itself, thus showing that the *35* can bootstrap itself, and that the forces due to 35-exchange are just zero in all other states so that we get a minimal bootstrap.

Starting with the reciprocal bootstrap of $N$ and $N^*$, and the associated condition $\Gamma = C\Gamma$, 'we are thus finally led to the modified bootstrap condition (22) in terms of which the *35* and *56* form a simple closed system. Moreover, this system includes all the low-lying meson and baryson states.

# REFERENCES

1. G. F. Chew and S. Mandelstam, *Nuovo Cimento* **19**: 752 (1961).
2. G. F. Chew, *Phys. Rev. Letters* **9**: 233 (1962).
3. E. Abers, L. A. P. Balázs, Y. Hara, *Phys. Rev.* **136**: B 1382 (1964).
4. G. Wentzel, *Rev. Mod. Phys.* **19**: 1 (1947).
5. T. Cook, C. J. Goebel, and B. Sakita, *Phys. Rev. Letters* **15**: 35 (1965).
6. V. Singh, *Phys. Rev.* **144**: 1275 (1966).
7. A. Kumar, *Phys. Rev.* **148**: 1347 (1966).
8. V. Singh and B. M. Udgaonkar, *Phys. Rev.* **149**: 1164 (1966); C. Goebel, Proceedings of the 12th International Conference on High Energy Physics, Vol. 1, Dubna, 1964, Atomizdat, Moscow, 1966, p. 255.
9. R. Dashen, *Phys. Letters* **11**: 89 (1964). Y. Hara, *Phys. Rev.* **135**: B 1079 (1964).
10. C. Goebel, Ref. 8. V. Singh, unpublished.
11. L. A. P. Balázs, V. Singh, and B. M. Udgaonkar, *Phys. Rev.* **139**: B 1313 (1965). R. C. Hwa and S. H. Patil, *Phys. Rev.* **138**: B 933 (1965).
12. R. H. Capps, *Phys. Rev. Letters*, **14**: 31 (1965); J. G. Belinfante and R. E. Cutkosky, *Phys. Rev. Letters* **14**: 33 (1965).

13. V. Singh and B. M. Udgaonkar, *Phys. Rev.* **139**: *B* 1585 (1965).
14. C. Goebel, *Phys. Rev. Letters* **16**: 1130 (1966). See also T. Cook and B. Sakita, *J. Math. Phys.* **8**: 708 (1967).
15. J. Kuriyan and E. C. G. Sudarshan, *Phys. Letters* **21**: 106 (1966); *Phys. Rev. Letters* **16**: 825 (1966).
16. J. Fulco and D. Y. Wong, *Phys. Rev. Letters* **15**: 274 (1965).
17. B. M. Udgaonkar, Institute for Advanced Study preprint, July, 1966.

# Quark Model for Baryons and Resonances

## A. N. MITRA[†]

*UNIVERSITY OF DELHI*
*Delhi, India*

---

I would like to present a quark model for the structure of baryons and their resonances which we have developed during the past few months and attempted to apply to a few physical processes like (*a*) strong decay of baryon resonances, (*b*) meson baryon scattering and (*c*) photoproduction of pseudoscalar mesons.

The idea that the structure of baryons and their resonances can be understood in terms of nonrelativistic quarks $Q$ is both physically interesting mathematically tractable.[1,2] Even without any detailed dynamical assumptions about the $Q$–$Q$ forces, a quark model already predicts several interesting $SU(6)$ results like the $F/D$ ratio, the ratio of the nucleon magnetic moments, $\omega$–$\phi$ mixing angle, and so on.[1,2] One, however, needs more detailed dynamical assumptions to *predict* the group structures of the various baryons, their central masses, modes of splitting due to spin-orbit, $SU(3)$-violating, etc., forces. For example, one might ask if there is a *dynamical* reason why the familiar octet and decuplet of baryons like to be grouped together into a **56** representation of $SU(6)$, in preference to others. Such attempts have been made, e.g., by Goebel[3] in the strong coupling theory with a good amount of success. We propose an approach in a similar spirit, but within the premises of the quark model.

---

†Department of Physics.

As for the basic assumptions, one must specify (*a*) the statistics for the quarks and (*b*) the nature of the forces. We are considering only the (most economical) quark model of Gell-Mann[4] and Zweig[5] (GMZ) rather than extended models[6-8] which are characterized either by more quarks or by additional quantum numbers. In the GMZ model there is a strong correlation between the representation of the $SU(6)$ group for the spin-cum-charge degrees of freedom and the spatial symmetry (*S*, *A*, or *M*) assumed for the $3Q$ wave function. Thus, Fermi statistics demands that the **56** representation must be associated with *A*-type spatial functions for the $3Q$ system. Unfortunately, *A*-functions have rather bad consequences for the structure of the baryon form factors.[9] Thus, it has been found that an *A*-function yields rather large oscillations in the form factor at values of $q^2$ as low as $\sim 20$ $F^{-2}$, in complete disagreement with experiment. On the other hand, an *S*-function predicts a smooth monotonic fall for large momentum transfers, which is at least in qualitative accord with observations. Such a big difference in the two predictions can be traced directly to the effect of the centrifugal barriers associated with *unit* magnitudes of the two independent angular momenta $l_1$ and $l_{23}$ within a $3Q$ system* which are essential for the construction of an *A*-function of $L = 0$ (*L* being the resultant *orbital* angular momentum of the three-body system). An *S*-function, on the other hand, does with zero values of $l_1$ and $l_{23}$, and thus avoids the centrifugal barriers. This difference suggests rather strongly that, at least for the usual baryons, *S*-functions must be preferred to *A*, and this in terms of statistics within the GMZ model implies a repudiation of Fermi statistics in favor of something which at least allows the appearence of *S*-functions in the **56**. The simplest possibility is perhaps to consider parastatistics which not only allows symmetrical wave functions but gives a kind of saturation at the level of three quarks.[10]

The second question concerns the nature of the quark forces. The simplest assumption is, of course, to invoke merely two-body forces, in keeping with conventional ideas. It is, of course, entirely possible that direct three-body forces are more important in this case, but it would perhaps be premature to invoke such an assumption even before some simple consequences of the two-body hypothesis have been explored. Now within the premises of $Q$–$Q$ forces,

---

*Indeed, if $l_{23}$ is the relative angular momentum of the (23) pair and $l_1$ the same for particle "1" with respect to the composite (23), then $L = l_1 + l_{23}$.

it can be shown[12] that an $S$-function of $L = 0$ which is most easily
generated by $s$-wave $Q$-$Q$ forces, has a strongly attractive kernel,
while an $A$-function of $L = 0$ which requires a $p$-wave $Q$-$Q$ interaction, has a repulsive kernel. This result, which has a fairly general
degree of validity, leads to a *dynamical* preference for parastatistics
over Fermi. Now, $p$-wave forces can be shown in a similar way
to predict a strongly attractive *axial vector*[11] state of $L^P = 1^+$, but
then such an $L$-value would not be compatible with the simple **56**
representation of $SU(6)$.

   With the two assumptions of parastatistics and two-body $Q$-$Q$
forces, the problem of baryon structures reduced essentially to a
quantum mechanical three-body problem. Moreover, within a
nonrelativistic model of quarks, one of our basic assumptions,[1,2] it
is only the *lower* partial waves in the two-body potentials that play
the essential role in the baryon structures, at least for the low lying
states. For this purpose, the $Q$-$Q$ potential could be so parameterized as to bring out only its lower partial waves, without reference to the higher ones. Such potentials which would necessarily
be nonlocal, could have the "radial parts" associated with each
partial wave parameterized in a simple form so as to make the $3Q$
problem mathematically tractable. The simplest assumption is the
factorable form which is known to reduce the three-body problem
to the calculational simplicity of a finite number of coupled
two-body problem.[12]

   We report here the results of this dynamical model characterized by parastatistics (which allows symmetrical wave functions)
and factorable $Q$-$Q$ forces in $s$- and $p$-waves, in which certain
bound states of $3Q$ are generated so as to exhibit characteristic
groups structures of $SU(6) \otimes O(3)$, thus providing a dynamical
realization of this group.[13] In the limit of "full symmetry" of the
$Q$-$Q$ interaction, which is analogous to a Wigner supermultiplet
potential,[14] the following states of $[SU(6) \otimes O(3)]^P$ appear in ascend-
ing order of energy levels:

$$(\mathbf{56}, 1)^+, (\mathbf{20}, 3)^+, (\mathbf{70}, 3)^-, (\mathbf{70}, 1)^+$$

While the **56** is, of course, realized as the state of lowest energy,
there appear some positive parity states before the convential naga-
tive parity ones,[2] namely, $(\mathbf{70}, 3)^-$, get their turn. We shall discuss
the implication of these positive parity states below.

   The spin and unitary-spin independent $Q$-$Q$ forces in $s$- and

$p$-states, consistent with parastatistics are of the form

$$V = (P_\sigma^+ P_u^+ + P_\sigma^- P_u^-) V_S + (P_\sigma^+ P_u^- + P_u^- P_u^+) \tag{2}$$

where $P_\sigma^\pm$ are the triplet and singlet spin-projection operators, and $P_u^\pm$ are the corresponding operators in unitary-spin space for the $2Q$ states of 6 and 3*, respectively.† With such a potential as equation (2), one obtains broadly the following structures for the $3Q$ states: The $s$-wave force $V_S$ yields $S$ or $M$ functions which satisfy Schrödinger equations of two distinct symmetry types. These equations, which are especially easy to handle with factorable shapes of $V_S$ and $V_p^{11}$ can be shown to be attractive for $S$, repulsive for $M$. This immediately yields the result that only the states (10, $S_{3/2}^+$) and (8, $S_{1/2}^+$) are strongly attractive, thus making up the desired 56, while remaining 70 states associated with $M$ functions are repulsive. The splitting into the multiplets 8 and 10 can be brought about through an $SU(3)$ invariant term of the form

$$(P_\sigma^+ P_u^+ + P_\sigma^- P_u^-) V_S' \tag{3}$$

and further splittings within each $SU(3)$ multiplet, by the usual $SU$(3)-violating terms. The $s$-wave part of equation (2) also yields a set of attractive states of $L^P = 1^-$ belonging to $(56, 3)^-$, but the kernel for these states are by themselves too weak to produce masses in the region of physical interest.

The $p$-wave part of (2) yields both $P = \pm$ states, The spatial functions are now of the type $A$ or $M$, but not $S$. For the negative parity states of $L^P = 1^-$, only the $M$-functions can be shown to have strongly attractive kernels, leading to the following multiplicity of states of (70, 3):

$$(10, {}^2p_{1/2}^-),\ (8,\ p_{1/2}^-)^2,\ (8,\ p_{3/2}^-)^2,\ (8, {}^4p_{5/2}^-),\ (1, {}^2p_{1/2}^-),$$
$$(1, {}^2p_{3/2}^-),\ (10, {}^2p_{3/2}^-) \tag{4}$$

For the $L^P = 1^+$ states, only the $A$-functions have strongly attractive kernels, yielding the following (20, 3)$^+$ states:

$$(1, {}^4P_{1/2}^+),\ (1, {}^4P_{3/2}^+),\ (1, {}^4P_{5/2}^+),,\ (8, {}^2P_{1/2}^+),\ (8, {}^2P_{3/2}^+) \tag{5}$$

There is a second set of $L^P = 0^+$ states which have attractive kernels for $M$, but not $A$. These are the following states of (70, 1)$^+$:

$$(10, S_{1/2}^+),\ (8, S_{1/2}^+),\ (8, S_{3/2}^+),\ (1, S_{1/2}^+) \tag{6}$$

---

†With Fermi statistics, the brackets multiplying $V_s$ and $V_p$ get interchanged.

Finally the $p$-wave interaction generate a set of positive parity states of $L^P = 2^+$ for which only $M$-functions have attractive, giving rise to the multiplicity $(70, 5)^+$ of $SU(6) \otimes O(3)$. These states are much higher in energy than those listed in (1), and are likely to be strongly affected by an (input) $d$-wave $Q$-$Q$ force.

These supermultiplets get split into different $SU(3)$ multiplets by a $p$-wave spin-orbit force of the (factorable) form:

$$M_Q\langle P|V_{LS}|P'\rangle = -3\lambda_{LS} P_\sigma^+ P_u^- i(\sigma_1 + \sigma_2) \mathbf{p} \times \mathbf{p'} v(p) v(p') \quad (7)$$

The appearence of $P_\sigma^+ P_u^-$ (associated) with the $p$-wave interaction under parastatistics) indicates that this force shifts only the 8 and 1 states, but not the 10. It also mixes the various $L$ states with the same $J^P$. If however this coupling is ignored, an extension of the calculational techniques in Ref. 11 leads to the following approximate formulas for the level splitting between different $SU(3)$ multiplets for the $(70, 3)^-$ and $(20, 3)^+$ states:

$$(70, 3)^-: \quad \Delta E^{(-)} \approx -B\xi \quad (8)$$

$$(20, 3)^+: \quad \Delta E^{(+)} \approx -B'\eta \quad (9)$$

where $B$ and $B'$ are certain parameters depending on the detailed dynamics and the numbers $\xi$ and $\eta$ are given according to the Table I. The $(70, 1)^+$ states are not split in this approximation.

As regards the experimental situation, the structure $(70, 3)^-$ of the negaitve parity states is qualitatively consistent with Dalitz analysis of the experimental data.[2] However, only for a few of the resonances the $J^P$ values seem to have been established with a reasonable amount of confidence. Even the $SU(3)$ assignments of these established resonances are in doubt. Thus, of the two $1/2^-$ octets and the $1/2^-$ decuplet only their N* components are known (1540, 1700, and 1680 MeV, respectively). Of the two $3/2^-$ octets and a $3/2^-$ decuplet predicted above, the known components are only N* (1518) Y*(1660) and $\Xi$*(1816). The known members of the $5/2^-$ octet are N*(1690) and Y*(1765). Finally, the two predicted singlets of $1/2^-$ and $3/2^-$ may be identified as Y*(1405) and Y*(1520), respectively, the higher mass of the latter being in accord with the prediction of Table I. It would of course be futile to attempt any detailed fit to these masses with so simple a model as the present one, yet the relative placings of some of the experimental resonances seem to be in qualitative accord with the values of the corresponding splitting parameters in Table I.

The most interesting prediction of the model is the natural appearance of certain energy states even below the negative energy ones. Thus after the spin-orbit corrections, Table I and equation (1) together predict a $1/2^+$ octet and a $1/2^+$ singlet to lie *below* the lowest octet and singlet among the negative parity particles. The N* member of the $1/2^+$ octet seems to fit rather well with the so-called Roper resonance at $\sim$ 1450 MeV, which has so far presented quite a challenge to any simple theoretical understanding on the quark model.[2] The physical reason for a natural appearence of such a state in our scheme is simply that an $A$-functions of $L^P = 1^+$ (axial vector) has a strongly attractive kernel.[11] This is a very general feature of the $p$-wave interaction, and the seprable approximation to the radial part does not affect its validity.

The model has a striking prediction of an $SU(3)$ singlet of positive parity at a very low mass [lower than Y*(1405)]. Such positive parity states would be much harder to detect than the conventional **56** of particles, because even their electromagnetic decay is greatly inhibited by strong selection rules[16] arising from the double recoil eflects which characterize their internal structures. One possibility to detect such a singlet could be to shoot a beam of $\Lambda$-particles against a proton target and look for a neutral object in an essentially forward direction.[17] Still another way could be to look at the strong decays of some of the negative parity baryons into these (positive parity) objects plus pseudoscalar mesons. This possibility, which is suggested on the basis of the result than some of these $L^P = 1^+$ states have lower masses than the $L^P = 1^-$ states, may be well worth while in view of the difficulty of production of such objects by any conventional mechanism (like one-pion excharge).

It may be remarked that parastatistics has played a rather helpful role in this investigation. In the **56**, it has allowed the use of $S$-functions which are not only essential for even a qualitative understanding of the baryon form factors, but also for a dynamical understanding of the strong $3Q$ binding which must go with the **56**. For the $L^P = 1^+$ states it makes a more economical prediction of a multiplicity $(\mathbf{20}, 3)^+$, as against the corresponding Fermi statistics result of $(\mathbf{56}, 3)^+$. The latter would have caused the embarrassing prediction of a (comparatively easily detectable) N*++ of $L^P = 1^+$ in the low mass region, as against a mere extra singlet of $L^P = 1^+$ under parastatistics. Finally, the spin-orbit term splits the singlets under parastatistics, decuplets under Fermi. Experimentally, the masses of the singlets Y*(1405) and Y*(1520) therefore have a more natural explanation under parastatistics, than under Fermi. Of course, any possible spltting in the decuplet must now come about through other symmetry violating mechanisms.

We have a!so applied this model to the calculation of certain processes which do not depend on the detailed dynamics or even the statistics (para or Fermi). The first application concerns the widths for the strong decays of the low-lying negative parity baryons into baryons plus pseudoscalar mesons.[18] The basic vertex for this purpose is the Yukawa-type coupling

$$Q' \rightarrow Q + \text{II}_8 \tag{10}$$

whose structure is determined on the basis of $SU(3)$ and Galilean invariance. This vertex is regarded as an operator whose matrix elements may be evaluated between appropriate $3Q$ states (initial and final). While the model gives standard $SU(6)$ results for the $\mathbf{56} \rightarrow \mathbf{56} + \text{II}$, it has several interesting results for the $(\mathbf{70}, 3)^-$ decays, going beyond $SU(3)$. An interesting result is that the spin-doublet member of the two $1/2^-$ octets is higher than the quartet member. A second important result is than the $\Xi^*(1816)$ is *not* a member of a $J^P = 3/2^-$ octet, rather the corrsponding decuplet.

A second application is in meson–baryon scattering and photo-production, through the use of the respective operators

$$Q + \text{II}_8 \rightarrow Q + \text{II}_8 \tag{11}$$

$$\mathfrak{r} + Q \rightarrow Q + \text{II}_8 \tag{12}$$

evaluated between appropriate $3Q$ states. For meson baryon scattering,[19] we obtain several $SU(6)$ sum rules for the familiar $\mathbf{56}$, like Johnson–Treiman[20] and Lipkin[21] relations, and in addition a few others which, though different from the predictions of $SU(6)$ or $SU(6)_W$, are not at least disfavored by experiment. The sum rules for the negative parity baryons are, of course, mere predictions at this stage. A simple variant of the operator (11), in which the initial pseudoscalar octet is replaced by a spurion octet, yields the known sum rules for nonleptonic decays of $\Sigma$, $\Lambda$, and $\Xi$, and a new one for $\Omega^-$ decay into $\Xi\pi$ and $\Xi^*\pi$. In a similar way, the photoproduction sum rules[22] obtained from (12) agree generally with $SU(6)$ results for the $\mathbf{56}$, and make several new predictions suitable for comparison with experiment.

In conclusion, I might remark that the quark model can be very useful as a tool for making predictions in the high-energy region using methods of low-energy physics. The interesting thing seems to be that these predictions are not more disfavored by experiment than those of more sophisticated theories.

### TABLE I
Spin-Orbit Parameters for Various $SU(3)$ States

| $L^P = 1^-$ states | $\xi$ | $L^P = 1^+$ states | $\eta$ |
|---|---|---|---|
| $(\mathbf{1}, 1/2^-)$ | 8/3 | $(\mathbf{1}, 1/2^+)$ | 5/3 |
| $(\mathbf{1}, 3/2^-)$ | $-4/3$ | $(\mathbf{1}, 3/2^+)$ | 2/3 |
| $(\mathbf{8}, 1/2^-)_\mathrm{I}$ | 4 | $(\mathbf{1}, 5/2^+)$ | $-4/3$ |
| $(\mathbf{8}, 1/2^-)_\mathrm{II}$ | $+2$ | $(\mathbf{8}, 1/2^+)$ | 4/3 |
| $(\mathbf{8}, 3/2^-)_\mathrm{I}$ | $-2$ | $(\mathbf{8}, 3/2^+)$ | $-2/3$ |
| $(\mathbf{8}, 3/2^-)_\mathrm{II}$ | 2 | | |
| $(\mathbf{8}, 5/2^-)$ | $-2$ | | |

## REFERENCES

1. G. Morpurgo, *Phys.* **2**: 95 (1965)
2. R. H. Dalitz, "Proceedings of the Oxford International Conference on Elementary Particles, 1965" (Rutherford High-Energy Laboratory, Harwell, England, 1966).
3. C. J. Goebel, *Phys. Rev. Letters* **16**: 1130 (1966).
4. M. Gell-Mann, *Phys. Letters* **8**: 214 (1964).
5. G. Zweig, CERN Report No. 8182/Th. 404 (1964).
6. M. Y. Han and Y. Nambu, *Phys. Rev.* **139**: B 1006 (1965).
7. H. Bacry, J. Nuyts, and L. Van Hove, *Phys. Letters* **9**: 279 (1964).
8. Y. Hara, *Phys. Rev.* **134B**: 701 (1964).
9. A. N. Mitra and R. Majumdar, *Phys. Rev.* **150**: 1194 (1966).
10. O. W. Greenberg, *Phys. Rev. Letters* **13**: 598 (1964).
11. A. N. Mitra, *Phys. Rev.* **142**: 1119 (1966).
12. A. N. Mitra, *Nucl. Phys.* **32**: 529 (1962).
13. K. T. Mahanthappa and E. C. G. Sudarshan, *Phys. Rev. Letters* **14**: 163 (1965).
14. E. P. Wigner, *Phys. Rev.* **51**: 105 (1937).
15. A. N. Mitra, *Phys. Rev.* **151**: 1168 (1966).
16. G. Morpurgo, *Phys. Letters* **20**: 684 (1966).
17. E. W. Anderson *et al. Phys. Rev. Letters* **16**: 105 (1966).
18. A. N. Mitra and M. H. Ross, *Phys. Rev.* **158**: 1630 (1967).
19. C. C. Joshi, V. S. Bhasin, and A. N. Mitra, *Phys. Rev.* **156** (5): 1572 (1967).
20. K. Johnson and S. B. Treiman, *Phys. Rev. Letters* **14**: 189 (1965).
21. H. J. Lipkin, *Phys. Rev. Letters* **16**: 1015 (1966).
22. S. Dass Gupta and A. N. Mitra, *Phys. Rev.* **156** (5): 1581 (1967).

# Riemann Mapping Theorem

## S. K. SINGH

*KARNATAK UNIVERSITY*
*Dharwar, India*

Riemann mapping theorem is one of the most important theorems in complex variables, and its proof utilizes almost all the important concepts of analytic functions. Before we state the theorem, let us recall some definitions. We shall use the symbol $\mathbb{C}$ for the set of complex numbers and $\mathbb{R}$ for the set of real numbers.

*Definition:* A continuous map $\gamma: I \to \mathbb{C}$ is called an *arc* or a *path* or a *curve* in $\mathbb{C}$, where $I = [0, 1]$ is the closed unit interval in $\mathbb{R}$. $\gamma(0)$ and $\gamma(1)$ are called the *initial* and *end* points of $\gamma$. If $\gamma(0) = \gamma(1) = z_0$ we say that $\gamma$ is a closed path *based* at $z_0$.

Two paths $\gamma_0, \gamma_1: I \to \mathbb{C}$ based at $z_0$ are said to be homotopic if there exists a continuous map $h: I \times I \to \mathbb{C}$ such that

$$h(t, 0) = \gamma_0(t) \quad \text{for } t \in I$$
$$h(t, 1) = \gamma_1(t) \quad \text{for } t \in I$$

and

$$h(0, u) = h(1, u) = z_0 \quad \text{for all } u \in I$$

In particular, we say that a closed path $\gamma_0$ based at $z_0$ is homotopic to a constant if $\gamma_0$ is homotopic to the constant map $\gamma_1: I \to \mathbb{C}$ defined by

$$\gamma_1(t) = z_0 \quad \text{for every } t \in I$$

**139**

*Definition:* A subset $A$ of $\mathbb{C}$ is said to be simply connected if every closed path $\gamma_0$ in $A$ based at $z_0$ is homotopic to the constant path $\gamma_1: I \to \mathbb{C}$ given by

$$\gamma_1(t) = z_0 \qquad \text{for every } t \in I$$

or, using the language of homotopy, if the fundamental group of $A$ is the identity group.

*Riemann mapping theorem:* If $D_1$ and $D_2$ are any two simply connected open subsets of $\mathbb{C}$ such that $D_1 \neq \mathbb{C}$, $D_2 \neq \mathbb{C}$, then $D_1$ and $D_2$ are isomorphic.

*Definition of isomorphism:* Let $A$ and $B$ be two non-empty open sets in $\mathbb{C}$. Then $A$ and $B$ are said to be isomorphic if there exists a map $f: A \to B$ such that $f$ is one-one, onto and both $f$ and $f^{-1}$ are analytic.

If $A$ and $B$ are isomorphic, we say that one is the conformal image of the other.

*Remark 1:* In the mapping theorem it is necessary that $D_1 \neq \mathbb{C}$ and $D_2 \neq \mathbb{C}$. For instance, if we take $D_1 = \mathbb{C}$ and $D_2 = \{|z| < 1\}$, then both $D_1$ and $D_2$ are simply connected open subsets of $\mathbb{C}$. But they are not isomorphic because if an isomorphism $f: \mathbb{C} \to \{|z| < 1\}$ exists, then $f$ is bounded and analytic in $\mathbb{C}$ and, hence, by Liouville's theorem, $f$ must reduce to a constant. But then $f$ will cease to be one-one.

Thus, we can say that $\mathbb{C}$ and $\{|z| < 1\}$ are not equivalent conformally, though topologically they are. We can set up a homeomorphism $g: \mathbb{C} \to \{|z| < 1\}$ given by

$$g(z) = \frac{z}{1 + |z|} \qquad \text{for } z \in \mathbb{C}$$

*Remark 2:* If $D_1$ is simply connected set and $D_2$ is not a simply connected set, then $D_1$ and $D_2$ are not isomorphic, because if $f: D_1 \to D_2$ is an isomorphism, then *a fortiori*, $f$ will be a homeomorphism and it is known that simply connectedness is a topological invariant.

Now if we pass from simply connected open subsets to open subsets of higher connectivity, the analogue of Riemann mapping theorem is no longer true. For instance the regions ($A$ nonvoid connected open subsets of $\mathbb{C}$ is called a region in $\mathbb{C}$)

$$\{1 < |z| < r\} \text{ and } \{1 < |z| < R\} \qquad (r \neq R)$$

are of the same connectivity, but they are not isomorphic. Suppose on the contrary that an isomorphism, $f: \{1 < |z| < r\} \to \{1 < |z| < R\}$ exists.

Let $f(z) = w$, then $h: \{1 < |z| < r\} \to \mathbb{R}$ defined by $h(z) = \log r \times \log |f(z)| - \log R \log|z|$ is harmonic in $1 < |z| < r$. Also $h(z)$ vanishes on the boundary, hence by the maximum principle $h(z) \equiv 0$.

Let $g: \{1<|z|<r\} \to \mathbb{C}$ be defined by $g(z) = \log r \log f(z) - \log R \times \log z$, then Re $\{g(z)\} = h(z)$. But $h(z) \equiv 0$; hence, Re $\{g(z)\} = 0$ so $g(z)$ must be an imaginary constant, say $g(z) = \log r \log f(z) - \log R \times \log z = i\alpha$. Now if $z$ describes the circle $|z| = r$ in the positive sense, $w = f(z)$ describes $|w| = R$ in the positive sense and both arg $(z)$ and arg $(f(z))$ grow by $2\pi$. But arg $(z) = $ Im $\{\log z\}$. Hence, imaginary part of $g(z)$ grows by the amount $2\pi (\log r - \log R)$ if $z$ describes $|z| = r$. But Im $\{g(z)\} = \alpha = $ constant. Hence, we must have $\log r - \log R = 0$. Thus $r = R$, this gives a contradiction (see Ref. 1, p. 333).

In order to prove Riemann mapping theorem it is sufficient to prove the following theorem:

*Theorem:* Any simply connected open subset $D$ of $\mathbb{C}$ ($D \neq \mathbb{C}$) is isomorphic to the unit disk $\{|z| < 1\}$.

Since if $D_1$ and $D_2$ are any two simply connected open subsets of $\mathbb{C}$ (both $\neq \mathbb{C}$) such that there exist isomorphisms

$$f: D_1 \to \{|z| < 1\}$$
$$g: D_2 \to \{|z| < 1\}$$

then the composed map $g^{-1} \circ f: D_1 \to D_2$ is an isomorphism.

As a matter of fact, with some conditions the isomorphism is unique. More precisely we state the final form of Riemann mapping theorem in the following way:

*Theorem:* Let $D$ be any simply connected open subset of $\mathbb{C}$ ($D \neq \mathbb{C}$); let $z_0$ be any given point of $D$. Then here exists a unique isomorphism $f: D \to \{|z| < 1\}$ such that $f(z_0) = 0$ and $f'(z_0) > 0$. The uniqueness of $f$ with the above conditions is easy to establish, i.e., suppose

$$f_1: D \to \{|z| < 1\}$$
$$f_2: D \to \{|z| < 1\}$$

are two isomorphisms such that

$$f_1(z_0) = 0 \qquad f_1'(z_0) > 0$$
$$f_2(z_0) = 0 \qquad f_2'(z_0) > 0$$

Then

$$f_1 \circ f_2^{-1}: \{|z| < 1\} \to \{|z| < 1\}$$

is an isomorphism and, hence, an automorphism of the unit disk $\{|z| < 1\}$. Let $g = f_1 \circ f_2^{-1}$, then $g(0) = (f_1 \circ f_2^{-1})(0) = f_1(f_2^{-1}(0))$ $= f_1(z_0) = 0$. Hence, $g$ is an automorphism of $\{|z| < 1\}$ which leaves 0 fixed. Hence, using Schwarz lemma it can be proved that $g(z) = \lambda z$ where $\lambda$ is a constant such that $|\lambda| = 1$. Hence,

$$g'(0) = \lambda$$

Also

$$g(z) = f_1(f_2^{-1}(z))$$

Hence,

$$g'(z) = f_1'(f_2^{-1}(z)) \frac{1}{f_2'(f_2^{-1}(z))}$$

so,

$$g'(0) = f_1'(z_0) \frac{1}{f_2'(z_0)} > 0$$

But

$$g'(0) = \lambda$$

Hence, $\lambda > 0$. Thus, since $|\lambda| = 1$, $\lambda = 1$. Hence, $g(z) \equiv z$ whenever $|z| < 1$. Thus, $g$ is an identity mapping. Hence, it follows that

$$f_1 = f_2$$

We shall not give the proof of the existence of the isomorphism $f$.

The Riemann mapping theorem shows that for all simply connected open subsets $D$ of $\mathbb{C}$ there exists a standard simply connected open set namely $\{|z| < 1\}$ onto which $D$ can be mapped conformally.

This analogue is not true for regions of connectivity $n$ ($n \geqslant 3$). That is, there does not exist a standard region onto which all regions of a given connectivity $n$ ($n \geqslant 3$) can be mapped conformally. Now let us point out why Riemann mapping theorem is so important. In tackling problems on boundary values, it happens many times that the boundary of the region is so complicated that it becomes impossible to get any decisive conclusion. But if the region happens to be simply connected (and not coinciding with the whole of $\mathbb{C}$) we can transform it conformally onto $\{|z| < 1\}$, and the properties of $\{|z| < 1\}$ are well known. The manipulations with $\{|z| < 1\}$ are easier.

Let us illustrate it with an examples. First, however, let us observe that Riemann mapping theorem throws no light on the behavior of the isomorphism $f$ on the boundaries of $D$ and that of $\{|z| < 1\}$. If we denote by $K$ the disk $\{|z| < 1\}$, it is not hard to construct a simply connected region $D$ such that the isomorphism $f$ is not continuous on the closure $\overline{D}$ of $D$.

However, if $\partial D$, the boundary of $D$ is a simple closed contour† then the isomorphism

$$f: D \to K \qquad K = \{|z| < 1\}$$

is continuous in $\overline{D}$ and for every point $\alpha$ such that when $|\alpha| = 1$, there exists one and only one $\beta \in \partial D$ such that $f(\beta) = \alpha$. Now let us see how the Riemann mapping theorem with this modification is used to find the solution of Dirichlet's boundary value problem.

*Dirichlet's boundary value problem:* Let $D$ be an open connected subset of $\mathbb{C}$. Let $\psi(z)$ be an arbitrary function defined on $\partial D$. Does there exist a function $u(\zeta)$ which is harmonic in $\overline{D}$ and such that $u|\partial D = \psi$?

If $D$ is any arbitrary open connected subset of $\mathbb{C}$, the solution of this problem is not known. However, if $D$ is a simply connected region $(D \neq \mathbb{C})$ such that $\partial D$ is a simple closed contour, then by the Riemann mapping theorem an isomorphism $f: \{|w| < 1\} \to D$ exists. Let $f(w) = z$, $w = u + iv$, and $z = x + iy = x(u, v) + iy(u, v)$. Let $\phi(x, y)$ be harmonic in $D$. Let $\psi(u, v) = \phi(x(u, v), y(u, v))$. Then $\psi(u, v)$ is harmonic in $|w| < 1$. Also by the Riemann mapping theorem in the modified form, there exists a one to one correspondance between the points $(x', y')$ of $\partial D$ and the points $(u', v')$ of $|w| = 1$. Thus, the boundary value of the harmonic function $\phi(x, y)$ at the point $(x', y')$ is the same as the boundary value of $\psi(u, v)$ at the corresponding point $(u', v')$ of $|w| = 1$. Thus, Dirichlet's problem amounts to its solution when $D$ is the unit disk $\{|z| < 1\}$, and this has a solution by simply applying Poisson's integral formula.

Using Riemann mapping theorem we can also find out the Green's function of a simply connected region $D$ $(\neq \mathbb{C})$ such that $\partial D$ is a simple closed contour.

---

†By a simple closed contour we mean a map $\gamma: [a, b] \to \mathbb{C}$ such that $\gamma(a) = \gamma(b)$ and $\gamma(t_1) = \gamma(t_2) \Rightarrow t_1 = a$ and $t_2 = b$, and such that it is possible to divide $[a, b]$ into a finite number of subintervals $I_1, \cdots, I_n$ such that $\gamma|I_K$ is differentiable and $\gamma'(t) \neq 0$ for $t \in I_K$. Instead of $[a, b]$ we could take $[0, 1]$.

We had so far been confined only to regions in $\mathbb{C}$. The theorem of Riemann is still true if we consider regions in $\overline{\mathbb{C}}$ where $\overline{\mathbb{C}} = \mathbb{C} \cup \{\infty\}$ is the usual one point compactification of $\mathbb{C}$. The neighborhood of $\infty$ is the complement of any compact subset of $\mathbb{C}$ to which $\infty$ has been added. But if we consider a simply connected region $D$ in $\overline{\mathbb{C}}$, then $\partial D$ must contain at least two boundary points.

Finally, let us point out that there is an alternative approach to the study of Riemann mapping theorem and also to the study of conformal mapping of multiply connected regions, through Bergman's theory of reproducing kernels. It has been proved by S. Bergman that if $D$ ($\neq \mathbb{C}$) is any simply connected open set in $\mathbb{C}$ and $z_0 \in D$, then given any $g(z) \in L^2(D)$, there exists a uniquely determined function $K(z, z_0) \in L^2(D)$ such that

$$g(z_0) = \iint_D g(z) \, \overline{K(z, z_0)} \, dx \, dy$$

where $K(z, z_0)$ is called Bergman reproducing kernel of $D$.

It can be proved that if $f(z)$ is the isomorphism given in Riemann mapping theorem (namely, the isomorphism $f: D \to \{|z| < 1\}$ is such that $f(z_0) = 0$ and $f'(z_0) > 0$, where $z_0 \in D$), then

$$f'(z) = \left( \frac{\pi}{K(z_0, z_0)} \right)^{1/2} K(z, z_0)$$

where $K(z, z_0)$ is the Bergman kernel function of $D$. (See Ref. 1, p. 252 and Ref. 2, p. 39).

## REFERENCES

1. Z. Nehari, "Conformal Mapping," McGraw-Hill Book Co., New York, 1952.
2. B. Epstein, "Orthogonal Families of Analytic Functions," Macmillan Co., New York, 1965.

# Field of Mikusinski Operators

## S. K. SINGH

*KARNATAK UNIVERSITY*
*Dharwar, India*

---

Let $\mathbb{C}$ be the set of complex numbers and $\mathbb{R}$ be the set of real numbers. Let

$$[0, \infty) = \{t \in \mathbb{R} \mid 0 \leqslant t < \infty\}$$

We shall denote by $C[0, \infty)$ the set of all mappings $a\colon [0, \infty) \to \mathbb{C}$ which are continuous. It is well known that if we define $a + b$ for $a, b \in C[0, \infty)$ by $(a + b)(t) = a(t) + b(t)$ for $t \in [0, \infty)$, then $(C[0, \infty), +)$ is an Abelian group. The "zero" of this group is the mapping $0\colon [0, \infty) \to \mathbb{C}$ defined by $0(t) = 0$ for every $t \in [0, \infty)$, and the negative $-a$ of $a$ is define by $(-a)(t) = -a(t)$ for $t \in [0, \infty)$. It is also known that if we define multiplication $\circ$ in $C[0, \infty)$ by $(a \circ b)(t) = a(t) \cdot b(t)$, then $C[0, \infty)$ is a ring with identity. It is not an integral domain, for instance, if

$$a(t) = 0 \qquad \text{for } 0 \leqslant t \leqslant 1$$
$$a(t) = t - 1 \qquad \text{for } 1 < t < \infty$$

and

$$b(t) = 1 - t \qquad \text{for } 0 \leqslant t \leqslant 1$$
$$b(t) = 0 \qquad \text{for } 1 < t < \infty$$

Then $a \neq 0$ and $b \neq 0$, but $a \circ b = 0$. Now our aim will be to define multiplication in $C[0, \infty)$ in such a way that $C[0, \infty)$ becomes an integral domain.

**145**

Let us define $a \cdot b$ for $a, b \in C[0, \infty)$ by $a \cdot b = (a \cdot b)(t) = \int_0^t a(t - u)b(u) \, du$ for $t \in [0, \infty)$. $a \cdot b$ is called the *convolution* (or faltung) of *a and b.* Since $\int_0^t a(t - u) \, b(u) \, du$ is a continuous function of $t$ in $[0, \infty)$ so $\cdot$ is really a mapping from $C[0, \infty) \times C[0, \infty) \rightarrow C[0, \infty)$. We can now verify that

$(i)$  $a \cdot b = b \cdot a$

$(ii)$  $(a \cdot b) \cdot c = a \cdot (b \cdot c)$

(This requires the change of order of integration in a double integral which is justifiable here.)

$(iii)$  $a \cdot (b + c) = (a \cdot b) + (a \cdot c)$

$(iv)$  There exists no $e \in C[0, \infty)$ with the property that $a \cdot e = e \cdot a = a$ for every $a \in C[0, \infty)$.

This is because if we let $l \colon [0, \infty) \rightarrow \mathbb{C}$ be defined by $l(t) = 1$ for $t \in [0, \infty)$, then $l$ is continuous, so $l \in C[0, \infty)$. Now if $a \cdot e = e \cdot a = a$ for every $a \in C[0, \infty)$, then $l \cdot e = (l \cdot e)(t) = \int_0^t l(t - y) \, e(y) \, dy = \int_0^t e(y) \, dy = l(t) = 1$. Thus $\int_0^t e(y) \, dy = 1$ for every $t \in [0, \infty)$ which is not possible, for instance, when $t = 0, \int_0^t e(y) \, dy = 0$. Hence, there exists no multiplicative identity. We shall write $ab$ for $a \cdot b$.

*Remark: l* defined above is called *Mikusinski* integral operator, because if $a \in C[0, \infty)$, then $l(a(t)) = (la)(t) = \int_0^t l(t - y) \, a(y) \, dy = \int_0^t a(y) \, dy$. Hence $l$ operating on $a$, integrates $a$.

$(v)$ Finally if $a, b \in C[0, \infty)$ such that $a \cdot b = 0$, then either $a = 0$ or $b = 0$; this is the most difficult part. The theorem is due to E. C. Titchmarsch. We leave the proof to the reader. We may indicate that the theorem is proved in two parts: first we assume that $a = b$. Then $a \cdot a = 0$ and we prove that $a = 0$. Then we use this part to prove finally that $a \cdot b = 0 \Rightarrow a = 0$ or $b = 0$. Collecting all these we get the following result:

*Theorem:* $(C[0, \infty), +, \cdot)$, where $\cdot$ is in the sense of convolution, is a commutative integral domain without multiplicative identity.

Now by the usual method we shall imbed this integral domain in a field (called its quotient field). The process is well known: We introduce an equivalence relation $\sim$ in $C[0, \infty) \times (C[0, \infty) - 0)$ as follows: $(a, b) \sim (c, d)$ iff $ad = bc$. As usual the equivalence class of $(a, b)$ will be written as $a/b$. If $F$ is the set of equivalence classes and if we define $+, \cdot$ in $F$ by

$$\frac{a}{b} + \frac{c}{d} = \frac{ad + bc}{bd}$$

$$\frac{a}{b} \cdot \frac{c}{d} = \frac{ac}{bd}$$

then $(F, +, \cdot)$ is a field, and the imbedding $\psi: C[0, \infty) \to F$ is given by

$$\psi(a) = \frac{ab}{b} \qquad b \neq 0$$

we usually ignore this imbedding and say that $C[0, \infty) \subset F$. The elements of $F$ are called *Mikusinski operators*. Let us note that if we write $1 = l/l$, then 1 is multiplicative identity of $F$. But $1 \notin C[0, \infty)$ since $C[0, \infty)$ has no identity. We can show that for every $\alpha \in \mathbb{C}$ there exists an operator $[\alpha] \in F$ and $\alpha \to [\alpha]$ gives the imbedding of $\mathbb{C}$ into $F$. Again ignoring the imbedding we can say that $\mathbb{C} \subset F$. One might ask that if instead of starting with $C[0, \infty)$ we had started with $L^1[0, \infty)$ functions, i.e., Lebesgue integrable functions complex valued and integrable over $[0, \infty)$, then we could have obtained its quotient field $F_1$. And if $a, b \in L^1[0, \infty)$, then $la, lb \in C[0, \infty)$ being continuous over $[0, \infty)$, and $a/b \to la/lb$ ($b \neq 0$) establishes an isomorphism between $F$ and $F_1$. Thus from purely algebraic point of view we have obtained nothing new. Thus it is desirable to start with $C[0, \infty)$ since it is easier to handle.

Now we show that the field $F$ of Mikusinski operators is not algebraically closed.

*Proof:* Consider the polynomial $z^2 = f$

where

$$f = (t \sin (\log t)) \qquad \text{for } t > 0$$

and

$$f(0) = 0$$

Thus $f \in C[0, \infty)$, hence $f \in F$ (by the imbedding). Hence, $z^2 - f$ is a polynomial with coefficients in $F$.

We shall say that the operator $x$ is real if $x$ can be represented by $p/q$ where $p, q$ are real valued continuous functions on $[0, \infty)$.

Then we can see that every operator $z$ can be expressed uniquly as $z = x + iy$ where $x, y$ are real operators. If we now suppose that $z^2 = f$ has a solution $z = x + iy$, then $x^2 - y^2 + 2ixy = f$. Hence, $x^2 - y^2 = f$ and $2xy = 0$. Hence, $x = 0$ or $y = 0$. Thus, if a solution exists, then it must be either purely real or purely imaginary. Now every operator $a$ is of the form

$$a = \frac{\{p(t)\}}{\{q(t)\}}$$

Define a transformation $U$ on operators by

$$U(a) = \frac{\{p(te^\pi)\}}{\{q(te^\pi)\}}$$

Then we can verify that

$$U(a_1 a_2) = U(a_1) U(a_2) \tag{1}$$

$$U(g(t)) = e^\pi \{g(te^\pi)\} \tag{2}$$

Then $z^2 = f$ given

$$U(z^2) = U(f)$$

$$U(z^2) = U(zz) = U(z)U(z) = U(f)$$

also

$$U(f) = U\{[t \sin (\log t)]\}$$
$$= e^\pi \{te^\pi \sin [\log (te^\pi)]\}$$
$$= - e^{2\pi}\{t \sin (\log t)\} = - e^{2\pi}f$$
$$= - e^{2\pi}z^2$$

Thus

$$U(z) = \pm ie^\pi z \tag{3}$$

But this is impossible, since $z$ and $U(z)$ are simultaneously real or imaginary. Equation (3), however, shows that if $U(z)$ is real, then $z$ is imaginary and if $U(z)$ is imaginary, then $z$ is real.

   The problem now arises as to how many elements need be adjuncted to the field $F$ of operators in order that any polynomial $p(x) \in F[x]$ may have a root in that adjuncted field. For example, for the field $\mathbb{R}$ of real numbers we know that the adjunction of a single element $i$ is sufficient. As a matter of fact, with the usual notation $\mathbb{R}(i) = \mathbb{C}$.

   The problem for the field of operators is an open problem. No one knows whether one, a finite number, countably infinite number, or a set of a given cardinality be adjuncted to $F$ to achieve the purpose.

Or let us take a simpler problem. A theorem, due to Kronecker, states that if $p(x)$ is a polynomial in a field $K$, there exists an extension $E$ of $K$ in which $p(x)$ has a root. We pose the following problem: What is the extension of $F$ in which the particular polynomial $z^2 - f$ has a root?

Or we can pose this problem: What is the splitting field of the above polynomial. The answer is not known. [It is known that if $p(x)$ is a polynomial in a field $K$, then there exists a splitting field $E$ of $p(x)$.] Let us recall: If $K \subset B \subset E$ are three fields we say that $B$ is *intermediate field*. If $E$ is an extension of $K$ in which the polynomial $p(x) \in K[x]$ can be factored into linear factors and if $p(x)$ cannot be so factored in any intermediate field, then we call $E$ the *splitting field* for $p(x)$.

Now we consider another problem of introducing topology in the field $F$ of operators.

*Definition:* Let $\{f_n\}_1^\infty$ denote a sequence in $C[0, \infty)$ We say that $\{f_n\}_1^\infty$ *converges almost uniformly* to a function $f \in C[0, \infty)$ if the sequence $\{f_n|[0, T]\}_1^\infty$ (restricted to $[0, T]$) converges uniformly to $f|[0, T]$ for every $T \in (0, \infty)$.

*Definition:* Let $F$ be the field of operators. Let $a_1, a_2, \ldots, a_n, \ldots$ be a sequence in $F$. We say that $\{a_n\}_1^\infty$ converges to $a \in F$ if there exists some $q \in F (q \neq 0)$ and a sequence $\{f_n\}_1^\infty$ in $C[0, \infty)$ converging almost uniformly to $f \in C[0, \infty)$ such that

$$\frac{a_n}{q} = \left(\frac{lf_n}{l}\right) \qquad \text{for } n = 1, 2, \ldots$$

and

$$\frac{a}{q} = \left(\frac{lf}{l}\right)$$

*Theorem:* If a sequence $\{a_n\}_1^\infty$ of operators in $F$ converges, then the limit is unique.

Hence, the above definition of convergence is reasonable. Now let us try to introduce a topology in $F$ in order that $F$ may be a topological field. A set $K$ furnished with a topology and field structure is called a topological field if the field operations are continuous in the topology.

Now suppose we want to introduce a topology in $F$ by the sequential convergence of operators defined above. With respect to this topology will $F$ be a topological field? This question has not been answered. Alternately let us introduce any topology in $F$ which renders $F$ as a topological field; the question is what kind of topological field $F$ will be. Let us recall the following theorem:

*Theorem:* If $K$ is a locally compact connected topological field with a countable base, then $K$ is isomorphic to either $\mathbb{R}$, $\mathbb{C}$, or $D$, where $\mathbb{R}$ is the field of real numbers, $\mathbb{C}$ is the field of complex numbers, and $D$ is the skew field of quaternions. $\mathbb{R}$, $\mathbb{C}$, and $D$ are taken with their usual field operations and their usual topologies. In $D$ we introduce norm by $\|x\| = \|a + bi + cj + dk\| = \sqrt{a^2 + b^2 + c^2 + d^2}$. (See Ref. 1, p. 173.)

We remark that two topological fields $K_1$ and $K_2$ are said to be isomorphic if there exists a map $f: K_1 \to K_2$ which is a homeomorphism of the topological space $K_1$ onto the topological space $K_2$ and which is at the same time an isomorphism of the field $K_1$ onto the field $K_2$.

If we now introduce any topology in the field $F$ of operators, such that $F$ becomes a topological field, then obviously $F$ will not be a locally compact connected topological field with a countable base because $F$ is neither isomorphic to $\mathbb{R}$ nor to $\mathbb{C}$ nor to $D$, because $\mathbb{C}$ is algebraically closed whereas $F$ is not.

Now, can we define a topology in $F$ such that $F$ becomes a locally compact (hence not connected) topological field? In that case how many components $F$ will have? Will it have a finite number of components? If not, what is the cardinality of the set of components? These questions are unanswered.

Finally, let us point out that many techniques of functional analysis could be used to study various properties of the field $F$ of operators, however we are confronted with a serious problem, namely we cannot introduce a norm in $F$ to give $F$ a Banach space structure in which convergence is equivalent to operational convergence (defined earlier). This can easily be demonstrated by proving the existence of a sequence $\{a_{mn}\}$ of operators in $F$ such that $\lim_{n \to \infty} a_{mn} = a_m$ (limit in operational sense) and $\lim_{m \to \infty} a_m = a$, but there exist no sequences of positive entegers $m_k$ and $n_k$ such that

$$\lim_{k \to \infty} a_{m_k \, n_k} = a$$

Whereas in any Banach space the above is true, namely, if $\{a_{mn}\}$ is a sequence in a Banach space $B$ such that $\lim_{n \to \infty} a_{mn} = a_m$

and $\lim_{m \to \infty} a_m = a$; then there exist sequences of positive integers $m_k$ and $n_k$ such that $\lim_{k \to \infty} a_{m_k \, n_k} = a$ (see Ref. 2, p. 363).

But the above theorem does not stop us in defining different kind of convergence in the field $F$ of operators, so that $F$ may acquire a Banach space structure.

Finally, let us point out that the field $F$ of operators has a lot of applications in the study of integral equations. Here we have studied the algebraic and topological aspect of it.

## ACKNOWLEDGMENTS

The author is thankful to Professor S. Drobot of Ohio State University for suggesting some of the problems raised in this article.

## REFERENCES

1. L. Pontrjagin, "Topological Groups," Princeton University Press, Princeton, New Jersey, 1958.
2. J. Mikusinski, "Operational Calculus," Pergamon Press, New York, 1959.

# On the Cosine Functional Equation

Pl. Kannappan

*UNIVERSITY OF WATERLOO*
*Waterloo, Canada*    and    *ANNAMALAI UNIVERSITY*
*Annamalainagar, India*

---

## 1. INTRODUCTION

Application of functional equations has preceded the development of a systematic theory of functional equations. One of the important applications of functional equations is a functional characterization of various functions like Euler's $\Gamma$ function, Lebesgue's singular function, cyclic functions, polynomials, exponential and logarithmic functions, etc. The most extensively studied problem of this sort is that of a functional characterization .of the trigonometric functions. One such example is the equation

$$f(x + y) + f(x - y) = 2f(x)f(y) \qquad \text{(A)}$$

The functional equation (A) has been extensively studied. The proof of the problem of parallelogram law of forces was reduced to the solution of (A) by D'Alembert.[3] Of course, the problem of parallelogram law of forces is one of the oldest problems solved by means of functional equations. This equation was considered for the same purpose by Poisson,[24] with a hypothesis of analyticity. On the line this equation (A) is satisfied by the cosine function and may be called the Cosine equation. In literature (A) is also known as D'Alembert's equation or Poisson equation. This equation (A) finds application in non-Euclidean mechanics and geometry.[2]

Cauchy[7] has proved that when $f$ is a real-valued continuous function of a real variable satisfying (A), then $f$ has one of the following forms: $f(x) \equiv 0$, $f(x) = \cos kx$, and $f(x) = (A^z + A^{-z})/2$, where $k$ is an arbitrary real number independent of $x$ and $A$ is an arbitrary positive constant. This result has been extended by S. Kaczmarz[14] to the case in which $f$ is a complex-valued, measurable function of a real variable (weaker hypothesis, measurability implies continuity). This result was further extended by T. M. Flett[8] as follows: Let $f(\not\equiv 0 \text{ or } 1)$ be a complex-valued function of a complex variable, continuous at at least one point and satisfy equation (A). Then, $f(z) = \cosh(\alpha x + \beta y)$, where $z = x + iy$, and $\alpha, \beta$ are complex numbers not both zero. This result was further extended[16] to the additive group $R^n$, where $R$ is the set of all real numbers and $n$ any positive integer, as follows: Let $f$ be a complex-valued function on $R^n$, continuous at some point and a solution of (A). If $f \not\equiv 0$, then $f(x) = \cosh(\alpha_1 x_1 + \alpha_2 x_2 + \cdots + \alpha_n x_n)$, where $x = (x_1, x_2, \ldots, x_n)$, $\alpha_i$'s are complex numbers. (See also Ref. 30.)

Andrade[4] gave the solution of (A) by assuming the continuity of $f$ over a certain interval, though integrability would suffice. Through integration he obtains the differentiability of $f$. Then he reduces (A) to the differential equation $f''(x) = cf(x)$ and obtains the solution. By using integral transforms,[2] one can also obtain the solution of (A). In these reductions, one has to assume much more than what is necessary and so one is bound to obtain naturally a weaker result. That is to say that the generality will be lost.

The equation (A) in abstract spaces (Hilbert spaces, Banach spaces, Banach Algebras, etc.) have been treated[20-22] by S. Kurepa and others. The following results in this direction are due to S. Kurepa[20]: Let $N(x)$ be a normal mapping of $R$ into $L(H)$ [where $L(H)$ is the set of all linear continuous mappings of a Hilbert space $H$ into $H$] which satisfies equation (A) for every $x$, $y$ in $R$. Suppose that $N(x)$ is a normal transformation for every $x$ in $R$, if $N(x) f = 0$, almost everywhere, then $f = 0$, and $N(x)$ is weakly continuous. Then a bounded self-adjoint transformation $B$ and a self-adjoint transformation $A$ exist such that $N(x) = [\exp(ixN) + \exp(-ixN)]/2 = \cos(xN)$ holds for all $x$ where $N = A + iB$.[21] Let $X$ be a Banach space and $f$ a complex-valued continuous functional defined on $X$, satisfying (A) with $f(0) = 1$. Then there exists an additive and continuous functional $a(x)$ such that $f(x) = \cos a(x)$, for $x$ in $X$. We can deduce from this the following result: Let $X$ be a Hilbert space and $f$ as above. Then $a(x)$ can be represented as $a(x) = \langle x, x_0 \rangle$, ($\langle \ \rangle$ denotes the inner product) where $x_0$ in $H$ is unique and independent of $x$. Hence, $f(x) = \cos \langle x, x_0 \rangle$.[21] Let $M$ denote the set of all real square matrices of order $n$ and $f$ a complex-valued, continuous functional defined on $M$ satisfying (A), with $f(0) = 1$ and $f(s^{-1}xs) = f(x)$, for each $x$ in $M$ and for every nonsingular matrix $s$ in $M$. Then $f(x) = \cos(a\mathrm{Tr}x)$, where $a$ is a complex number independent of $x$ and $\mathrm{Tr}\ x$ is the trace of the matrix $x$.

Problems similar to those in Ref. 11 gave rise to the following results[22]: Let $R$ be the real numbers, $B$ a real or complex Banach algebra with unit $e$ and $f\colon R \to B$, a measurable function satisfying (A) with $f(0) = e$. Then there is one and only one element $a$ in $B$ such that $f(t) = e + at^2/2! + a^2t^4/4! + \cdots$, the series being absolutely convergent for every $t$ in $R$.[22] Let $R$ be the reals, $X$ a Banach space, $B$ a Banach algebra with unit $e$, $F\colon X \to B$, the function satisfying (A) with $F(0) = e$, and $F$ is measurable on every ray, i.e., $F(tx)$ is measurable as a function of $t$ in $R$ for every $x$ in $X$. Then there exists a function $A(x)\colon X \to B$, such that $F(x) = e + A(x)/2! + [A(x)]^2/4! + [A(x)]^3/6! + \cdots$ and $A(x + y) + A(x - y) = 2A(x) + 2A(y)$. If $\|e - A(x_0)\| < 1$ for at least one $x_0$ in $X$, then an additive function $L\colon X \to B$ exists such that $A(x) = L(x)^2$ for every $x$ in $X$ and $F(x) = \cosh L(x)$. If $A(x)$ is continuous, then so is $L(x)$.

## 2. COSINE EQUATION ON GROUPS

Certainly the functional equation (A) has a meaning on any group $G$. We take $G$ to be a multiplicative group. Then (A) is of the form

$$f(xy) + f(xy^{-1}) = 2f(x)f(y) \tag{A}$$

One obvious way to solve the equation (A) is by means of a homomorphism of $G$, say $g$, into the multiplicative group of non-zero complex numbers $K$. If $g$ is such a homomorphism, then the function defined by

$$f(x) = \frac{g(x) + g^*(x)}{2} \tag{B}$$

for each $x$ in $G$ is a solution of (A), as can be seen by an easy calculation, where $g^*(x) = g(x)^{-1}$. Naturally, one would like to find out whether or not every solution of (A) on an arbitrary group has the form (B). If it is so, when $G$ is a topolopical group and $f$ is continuous, is $g$ also continuous? The answer to these questions are provided in the following sequence of results. (Note: for all the following see Ref. 15.)

*Lemma 1:* Let $G$ be any group. Let $g_1$ and $g_2$ be homomorphisms on $G$ into $K$ such that $g_1 + g_1{}^* = g_2 + g_2{}^*$. Then $g_2 = g_1$ or $g_1{}^*$.

*Lemma 2:* Let $G$ be an arbitrary Abelian group. Let $f$ be a function on $G$ with the properties that (1) $f$ satisfies (A) on $G$, (2) $f(x)$ assumes the values $\pm 1$ only on $G$. Then $f$ is of the form (B).

*Lemma 3:* Let $G$ be any cyclic group (finite or infinite). Then every solution of (A) on $G$ is of the form (B).

*Lemma 4:* Let $G_1$ and $G_2$ be two Abelian groups and that every solution of (A) on $G_i$ $(i = 1, 2)$ has the form (B). Then the same is true of all the solutions of (A) on the direct product $G_1 \times G_2$.

*Lemma 5:* Let $G_L (L \in I)$ be a family of Abelian groups such that every solution of (A) on $G_L$ has the form (B). Then the same is true of the weak direct product $P^*G_L$ ($x$ in $P^*G_L$ means $x = (x_L)$ and $x_L = e_L$ for all but a finite set of indices).

*Lemma 6:* Let $G$ be an arbitrary group on which all solutions of (A) are of the form (B). Let $H$ be a homomophic image of $G$. Then every solution of (A) on $H$ has the form (B).

Piecing together the above results we can prove the following theorem:

*Theorem 1:* Let $G$ be any Abelian group. Then every solution of (A) on $G$ has the form (B).

*Proof:* It is well known that every Abelian group $G$ with $m$ generators is a homomorphic image of the weak direct product $Z^{m^*}$, $Z$ being an infinite cyclic group. But every solution of (A) on $Z^{m^*}$ is of the form (B) by Lemmas 3 and 5. Now an application of Lemma 6 shows that all solutions of (A) on $G$ are of the form (B). Since there is no restriction whatsoever on the cardinality $m$ for the validity of Lemma 5, we conclude that every function that satisfies (A) on an arbitrary Abelian group $G$ has the form (B).

In case $G$ is a non-Abelian group, let $f$ satisfy

$$f(xyz) = f(xzy) \tag{C}$$

for all $x, y, z$ in $G$. Then every solution of (A) on $G$ satisfying (C) has the form (B). (See Ref. 17.)

*Theorem 2:* Let $G$ be a topological group (Abelian or not). Suppose $g: G \to K$ is a homomorphism and $f(x) = [g(x) + g^*(x)]/2$ is continuous. Then $g$ is continuous.

*Proof:* Let $\{x_\delta\}$ be a net in $G$ converging to the element $x_0$. We must show that $\lim g(x_\delta) = g(x_0)$. The net $\{g(x_\delta)\}$ of complex numbers is bounded. For, if there were a subnet $\{g(x_\alpha)\}$ with $|g(x_\alpha)| \to \infty$, then $|g(x_\alpha)|^{-1} \to 0$. Hence, $2f(x_\alpha) - g(x_\alpha) = 1/g(x_\alpha) \to 0$, a contradiction to the fact that $\lim f(x_\alpha) = f(x_0)$ is finite. Similarly, the net $\{1/g(x_\delta)\}$ is bounded [look at $2f(x_\alpha) - 1/g(x_\alpha) = g(x_\alpha)$]. That is, $\{g(x_\delta)\}$ is bounded away from zero. Now we will show that *any* subnet $\{x_\alpha\}$ of $\{x_\delta\}$ for which $\{g(x_\alpha)\}$ converges to a non-zero complex number has the property that $\lim g(x_\alpha) = g(x_0)$. Let $a = \lim g(x_\alpha)$ and $g(x_0)$ $= b$. Then $(b + b^{-1})/2 = f(x_0) = \lim f(x_\alpha) = \lim [g(x_\alpha) + g(x_\alpha)^{-1}]/2$ $= (a + a^{-1})/2$. Hence, $b - a = (b - a)/ab$. Thus either $b = a$ or $b = 1/a$. If $b = 1/a$, we will show that $1/a = a$, so that in either case $b = a$. Indeed we know that $f(x^2) + 1 = 2f(x)^2$ for every $x$ in $G$. Hence, $f(x_0^2) = 2f(x_0)^2 - 1 = [2(b + b^{-1})^2/4] - 1 = 1/2\,(b^2 + b^{-2})$. On the other hand, $x_\alpha \to x_0$ implies that $x_\alpha x_0 \to x_0^2$. So, $f(x_0^2) = \lim f(x_\alpha x_0)$ $= \lim [g(x_\alpha x_0) + g(x_\alpha x_0)^{-1}]/2 = (ba + 1/ba)/2 = 1$(since $b = 1/a$). So, $2 = b^2 + 1/b^2 = a^2 + 1/a^2$. Hence, $(a^2 - 1)^2 = 0$. Therefore, $a^2 = 1$ or $a = 1/a = b$. This implies that $g$ is continuous. For if $\lim g(x_\delta)$ $\neq g(k_0)$, there would exist a subnet $\{x_\beta\}$ of $\{x_\delta\}$ with $|g(x_\beta) - g(x_0)|$ $\geqslant \epsilon > 0$ and a subnet $\{x_\alpha\}$ of $\{x_\beta\}$ which converges to a non-zero number. Again we have a contradiction. Hence, $g$ is continuous. The proof given here is due to Prof. R. R. Phelphs. The following theorem is now obvious.

*Theorem 3:* Let $G$ be a topological group. Let $f$ be a continuous solution of (A) on $G$ satisfying (C). Then, $f$ is of the form (B), where $g$ is a continuons homomorphism of $G$ into $K$.

Suppose, in Theorem 3, $G$ is locally compact and $f$ is (Haar) measurable. Then, it is easy to see that $g$ is also measurable (See Ref. 17). But since $g$ is a measurable homomorphism, $g$ is also continuous. Hence, $f$ is continuous. So, every measurable solution of (A) is continuous.

## 3.  DISCUSSION

Therefore, a function $f$ on $G$ satisfying (C) satisfies (A) if and only if $f$ is of the form (B). Now, $g$ can be written in one and only one way in the form $g = \chi \exp(\varphi)$, where $\chi$ is a character of $G$,[10] and $\varphi$ is a homomorphism of $G$ into the additive reals. Further, if $g$ is continuous, then so are $\chi$ and $\varphi$.

Indeed, since $g$ is a homomorphism into $K$, $g/|g|$ is a character of $G$, say $\chi$. If $g$ is continuous, then so is $\chi$. Now let $\varphi = \log |g|$. Then plainly enough $\varphi$ is a homomorphism of $G$ into the additive reals, called a real character,[10] and $|g| = \exp(\varphi)$. Further, it is evident that $\varphi$ is continuous when $g$ is. Hence, we have $g = \chi \exp(\varphi)$. So, computing $g$'s (or $f$'s) is reduced to computing characters and real characters of $G$.

*Example:* Let $1 \leq n < \infty$. Let $f$ be a continuous solution of (A) on $R^n$, where $R^n$ is a topological space as a Cartesian product of the reals $R$. By Lemma 4, it is enough to know the homomorphisms on the component groups, that is here on $R$ with the usual topology (see Ref. 15). So, if $f$ is a continuous solution of (A) on $R$, then $f/Q$ is also a continuous solution of (A) on $Q$, where $Q$ is the additive group of rational numbers. The continuity of $f$ ensures the continuity of the corresponding $g$ on $Q$ satisfying (B). Further $g(x) = g(1)^x$, $x$ in $Q$. So, $|g(x)| = |g(1)|^x = \exp(\alpha_x)$, where $\alpha = \log |g(1)|$ [$|g|$ is real and greater than zero]. Now, $g/|g|$ is a continuous character of $Q$ and, therefore, has the form $\exp(i\beta x)$, $\beta$ real (see Ref. 10, p. 414). So, $g(x) = \exp[(\alpha + i\beta)x]$ on $Q$. $f = (g + g^*)/2$ on $Q$ and therefore on $R$. Hence, all continuous solutions $f$ of (A) on $R$ have the form $f(x) = \cosh(\alpha x)$, where $\alpha$ is any complex number. Now applying Lemma 4, we conclude that every continuous solution of (A) on $R^n$ is given by

$$f(x_1, x_2, \ldots, x_n) = \frac{g(x_1, x_2, \ldots, x_n) + g^*(x_1, x_2, \ldots, x_n)}{2}$$

where

$$g(x_1, x_2, \ldots, x_n) = \prod_{p=1}^{n} \exp[(\alpha_p + i\beta_p)x_p]$$

$$= \exp(\sum_{p=1}^{n} [\alpha_p + i\beta_p)x_p]$$

So, $f(x_1, x_2, \ldots, x_n) = \cosh[\sum_{p=1}^{n}(\alpha_p + i\beta_p)x_p]$. This generalizes Flett's and earlier results.[4,7,8,14]

There do exist unbounded, discontinuous solutions of (A). Consider the $p$-adic integers, $\triangle_p$.[10] Now $\triangle_p \supset Z^{c*}$, where $c$ is the continuum. Since there are many unbounded, discontinuous real characters of $Z$ and each one of it can be extended to a real character on $\triangle_p$, it is evident that there are many unbounded, discontinuous solutions of (A) on $\triangle_p$.

## 4.  SOME GENERALIZATIONS OF THE EQUATION (A)

The functional equation

$$f(x + y) + f(x - y) = 2f(x)\cos y \qquad (4.1)$$

is more akin to (A). By simple substitution one obtains the solution of (4.1) as $f(x) = a \cos x + b \sin x$, (see Ref. 2). Wilson's first generalization of (A) (see Refs. 29 and 30), the functional equation

$$f(x + y) + f(x - y) = 2f(x)g(y) \qquad (4.2)$$

is certainly a generalization of (A). Wilson's method of solving this equation (4.2) is based on decomposing $f$ into an even and odd components and obtaining the functional equation (A) satisfied by $g$. The most general continuous solutions of (4.2) are $f \equiv 0$, $g$ arbitrary or $f(x) = c \cos bx + d \sin bx$, $g(x) = \cos bx$ or $f(x) = \cosh bx + d \sinh bx$, $g(x) = c \cosh bx$ or $f(x) = c + dx$, $g \equiv 1$. Wilson's second generalization of (A) (see Refs. 29 and 30), the functional equation

$$f(x + y) + g(x - y) = h(x)k(y) \qquad (4.3)$$

can also be considered as a generalization of (A). Here again Wilson used the technique of splitting the function into an even and odd components and reducing it into the form (4.2). The functional equation $f(x + y) = F[f(x - y), f(x), f(y), x, y]$, is a generalization of (A). For a discussion of it see Ref. 2.

## 5.  CHARACTERIZATION OF THE COSINE[18]

Let $f$ and $g$ be functions satisfying

$$f(x - y) = f(x)f(y) + g(x)g(y) \qquad (5.1)$$

and

$$\lim_{x \to 0} g(x)/x = 1 \qquad (5.2)$$

Then $f(x) = \cos x$ and $g(x) = \sin x$. If, however, one wants to define cosine alone, one may use the equation

$$f(x + y) = f(x)f(y) - \sqrt{1 - f(x)^2}\,\sqrt{1 - f(y)^2} \qquad (5.3)$$

whose general continuous solution is $f(x) = \cos kx$.[9,2] Equation (5.1) or (5.3) certainly characterizes the cosine function. But it would perhaps be desirable and at least interesting to characterize the cosine by an equation in a single variable. In this direction, the following functional equations

$$f(2x) = 2f(x)^2 - 1 \tag{5.4}$$

and

$$f(x/2) = s(x) \sqrt{[1 + f(x)]/2} \tag{5.5}$$

where

$$s(x) = \begin{cases} 1 & \text{for } x \in [(4k-1)\pi, (4k+1)\pi[ \\ -1 & \text{for } x \in [(4k+1)\pi, (4k+3)\pi[ \ k = 0, \pm 1, \pm, 2 \end{cases}$$

can be considered.

It has been proved that $f(x) = \cos x$ is the only even solution of (5.4) which is twice differentiable at $x = 0$ and such that $f(0) = 1$ and $f''(0) = -1$. Also $f(x) = \cos x$ is the only function defined for all $x$, continuous in a neighborhood of $x = 0$, satisfying (5.5) and periodic with period $2\pi$. These results characterize the cosine. In (5.4), it would be nicer to be able to introduce the cosine before introducing the derivatives, with no stronger means than, e.g., continuity. But it has been proved that the condition that $f$ is twice differentiable cannot be weakened, even under the additional assumption of continuity. In (5.5), it contains the sign factor which is somewhat disagreeable. So, the natural question arises, whether the cosine cannot be defined with the aid of (5.4), periodicity and continuity. So, the problem of a characterization of the cosine by a functional equation in a single variable remains unanswered.

## 6. ON SOME FUNCTIONAL INEQUALITIES CONCERNING THE EQUATION (A)

First, consider the functional inequality

$$|f(x+y)f(x-y)| \le |f(x)|^2 + |f(y)|^2 \tag{6.1}$$

It has been proved[12] that if $f(x)$ is an entire function of $x$, then all the functions which satisfy (6.1) are $\exp(\alpha x + \beta)$, $\alpha \sin \beta x$, $\alpha x$ where $\alpha, \beta$ are arbitrary complex constants, and only these. Next, let us consider the functional inequality

$$2|f(x)f(y)| \le |f(0)| \, (|f(x+y)| + |f(x-y)|) \tag{6.2}$$

It has been proved[12] that if $f(x)$ is an entire function of $x$, then the functions which satisfy (6.2) are $\alpha \cos^n \beta x$, $\alpha \exp(\beta x^2)$ where $\alpha$, $\beta$ are arbitrary complex constants and $n$ is an arbitrary natural number, and only these. Finally, let us consider the functional inequality

$$2|f(x)g(y)| \leq |f(x+y)| + |f(x-y)| \qquad (6.3)$$

It has been proved that if $f(x)$, $g(x)$ are entire functions of $x$ with $g(0) = 1$ and satisfying the functional inequality (6.3), then $f(x) = a \cos^n(bx + c)$ or $f(x) = \exp(ax^2 + bx + c)$ or $f(x) = (ax + b)^n$ where $a, b, c$ are arbitrary complex numbers and $n$ is an arbitrary natural number. The solution of (6.3) are only these. One can also solve the functional equation (A) for complex variables with the help of either (6.1), (6.2), or (6.3).

## REFERENCES

1. J. Aczel, "Fuggvenyegyenletek az alkalmazott matematikaban." *Magyar Tud. Akad. Mat. Fiz. Oszt. Kozl* **1**: 131–142 (1951).
2. J. Aczel, "Lectures on Functional Equations and Their Applications," Acadamic Press, 1966.
3. D'Alembert, "Memoire sur les principes de mecanique," *Hist. of Acad. Sci.* Paris, 278–286 (1769).
4. J. Andrade, "Sur l'equation fonctionnelle de Poisson," *Bull. Soc. Math. France,* 58–63 (1900).
5. G. Arrighi, "Sulla equazione funzionale $2\varphi(x)\varphi(y) = \varphi(x+y) + \varphi(x-y)$," *Boll. Unione Mat. Ital.,* **4**: 255–257 (1949).
6. R. D. Carmichael, "On Certain Functional Equations," *Amer. Math. Monthly,* **16**: 58–63 (1909).
7. A. L. Cauchy, "Ouevre Completes, II series," T III: Paris, 98–113 (1897).
8. T. M. Flett, "Continuous Solutions of the Functional Equation $f(x+y) + f(x-y) = 2f(x)f(y)$," *Amer. Math. Monthly,* **70**: 392–397 (1963).
9. M. Ghermanescu, "Sur la definition fonctionnelle des fonctions trigonometrique," *Publ. Math. Debrecen* **5**: 93–96 (1957).
10. E. Hewitt and K. A. Ross, "Abstract Harmonic Analysis," Springer Verlage, 1963.
11. E. Hille and R. S. Phillips, "Functional Analysis and Semi-Groups," *Amer. Math. Soc.,* Colloquium Publ. **31**: (1957).
12. Hiroshi Haruki, "Studies on Certain Functional Equations from the Standpoint of Analytic Function Theory," *Osaka Univ, Sci. Reports,* **14**: 1–40 (1965).
13. D. V. Ionescu, "Quelques applications de certaines équations fonctionnelle," *Mathematica (Timisora)* **19**: 159–166 (1943).
14. S. Kaczmarz, "Sur l'equation fonctionnelle $f(x) + f(x-y) = \varphi(y)f(x+y/2)$" *Fund. Math.,* **6** 122–129 (1924).
15. Pl. Kannappan, "The Functional Equation $f(xy) + f(xy^{-1}) = 2f(x)f(y)$ for Abelian Groups," Ph. D. thesis, Univ. of Washington, Seattle, 1964.

16. Pl. Kannappan, "On the Functional Equation $f(x + y) + (x - y) = 2 f(x) f(y)$," *Amer. Math. Monthly* **72**: 374–377 (1965).

17. Pl. Kannappan, "Functional Equation $f(xy) + f(xy^{-1}) = 2f(x)f(y)$ for Groups," *Proc. Amer. Math. Soc.* **19**(1): 69(1967).

18. M. Kuczma, "On a Characterization of the Cosine," *Ann. Polon. Math.* **16**: 53–57 (1965).

19. S. Kurepa, "A Cosine Functional Equation in $n$-Dimensional Vector Space," *Glasnik Mat. Fiz. Astronom.* **14**: 169–189 (1959).

20. S. Kurepa, "A Cosine Functional Équation in Hilbert Spaces," *Cand. J. Math.* **12**: 45–50 (1960).

21. S. Kurepa, "On Some Functional Equation in Banach Spaces," *Studia Math.* 149–158 (1960).

22. S. Kurepa, "A Cosine Functional Equation in Banach Algebras," *Acta Sci. Math. Sceged,* **23**: 255–267 (1962).

23. G. Maltese, "Spectral Representations for Solutions of Certain Abstract Functional Equations," *Composito Math.* **15**: 1–22 (1961).

24. S. D. Poisson, "Du parellelogramme des forces" Correspondance sur l'École Polytechnique, 356–360 (1804).

25. R. Shimmack, "Axiomatische Untersuchungen über die Vektroaddition," Dissertation, Halle, 1908.

26. R. Shimmack, "Axiomatische Untersuchungen über die Vektoraddition," *Abhandl. Akad. Naturforsch.* Halle **90**: (1909).

27. Van Der Lyn, "Sur l'equation fonctionnelle $f(x + y) + f(x - y) = 2 f(x) \varphi(y)$," *Math. Cluj* **16**: 91–96 (1940).

28. H. E. Vaughan, "Characterization of the Sine and Cosine," *Amer. Math. Monthly* **62**: 707–713 (1955).

29. W. H. Wilson, "Two General Functional Equations," *Bull. Amer. Math. Soc.* **31**: 330–334 (1925).

30. W. H. Wilson, "On Certain Related Functional Equations," *Bull. Amer. Math. Soc.* **26**: 300–312 (1919–20).

# A Novel Approach to the Kinetic Theory of Fluids—Onset of Turbulent Motion

S. K. SRINIVASAN

*INDIAN INSTITUTE OF TECHNOLOGY*†
*Madras, India*

---

## 1. INTRODUCTION

It is well known that the macroscopic properties of a fluid in a state of motion are described by the classical equations of hydro-dynamics due to Euler, Lagrange, and Stokes. The equations that have been "derived" by an ingenious application of Newton's laws of motion to an infinitesimally small volume of fluid have found a variety of applications to specific physical or engineering situations. For instance, in the case of the flow pattern around some body or the flow around a high-speed projectile and in many other problems of a highly intricate nature have been studied by many workers in this field. While problems of this type based on the original equation of Euler and others are reaching a comparatively saturated state, the elucidation of the physical basis of the equations of hydrodynamics has not received much attention until quite recently. Only in the past few years some attempts have been made toward a deduction of the laws of fluid motion from the results of the kinetic theory of fluids.[1] Encouraged by such attempts, the author[2] has attempted to

---

†Department of Mathematics.

derive *macroscopic* properties with the help of certain correlation functions called product densities which are familiar in the theory of stochastic point processes.[3] The hydrodynamical equations that follow from the heirarchy of equations satisfied by the product densities have been shown to be valid only for configurations very near the equilibrium position determined by the Poisson approximation. On the other hand, for configurations which are far from equilibrium position where the Poisson approximation ceases to be valid, we have demonstrated, though qualitatively, the onset of instability arising from the violation of the usual fluctuation theorem of statistical mechanics. In the present contribution, we wish to amplify the qualitative arguments by quantitaitve estimates of the fluctuation of physically measurable quantities by seeking small perturbations about the mean position which is normally stable for systems at local equilibrium. Such a method of approach enables us to conclude that the macroscopic properties of the fluid, in spite of their deterministic behavior to start with, do not continue to obey the Navier–Stokes equation. It appears that we may have to introduce two new features into the Navier–Stokes equation in order to predict the macroscopic properties of the fluid. One of them consists in the introduction of some kind of a *stirring force* which is random in character and arises from the fluctuation of the force due to some kind of a Brownian motion suffered by a molecule due to the presence of other molecules. In fact this feature readily explains the unproved hypothesis of the stirring force which was put forward by Edwards[4] and subsequently developed by Kraichnan.[5] The other noteworthy feature that emerges leads us to the conclusion that the response coefficients are defined only statistically. It is exactly for this reason that we experience difficulty in interpreting the coefficient of viscosity and thermal conductivity in configurations far from local equilibrium. In view of the difficulty in obtaining an explicit nonperturbative solution to the generalized Boltzmann equations, we take an alternative but useful approach to the description of the hydrodynamical system by incorporating in the Navier–Stokes equations appropriate parameters and functions to account for large deviations from the mean values. In the final section, we shall deal with a linearized equation which explains the turbulent motion of compressible fluids when the Reynolds number is in a certain range and the consequent production of resonating sound waves, a phenomenon observed to be characteristic of air turbulence.

## 2. GENERALIZED BOLTZMANN EQUATIONS

Let $dN(\mathbf{p}, \mathbf{q}; t)$ be the number of molecules that are found at time $t$ with momentum between $\mathbf{p}$ and $\mathbf{p} + d\mathbf{p}$ and coordinate between $\mathbf{q}$ and $\mathbf{q} + d\mathbf{q}$. If we adopt the Gibbsian ensemble approach, it is clear that $dN(\mathbf{p}, \mathbf{q}; t)$ is a random variable and we encounter the problem of the distribution of a discrete number of molecules distributed over a continuum of the phase space. Thus if we use the methods of stochastic point processes, we can introduce the functions $f_1(\mathbf{p}, \mathbf{q}; t), f_2(\mathbf{p}_1, \mathbf{q}_1; \mathbf{p}_2, \mathbf{q}_2; t), \cdots$ where (see Ramakrishnan[6])

$$f_1(\mathbf{p}, \mathbf{q}; t) \, d\Omega_1 = \overline{dN(\mathbf{p}, \mathbf{q}; t)}$$

$$f_2(\mathbf{p}_1, \mathbf{q}_1; \mathbf{p}_2, \mathbf{q}_2; t) \, d\Omega_1 \, d\Omega_2 = \overline{dN(\mathbf{p}_1, \mathbf{q}_1; t) \, dN(\mathbf{p}_2, \mathbf{q}_2; t)}$$

$$\cdots \quad \cdots \quad \cdots \quad \cdots \quad \cdots \quad \cdots \quad \cdots \quad \cdots$$

$$\cdots \quad \cdots \quad \cdots \quad \cdots \quad \cdots \quad \cdots \quad \cdots \quad \cdots$$

$$f_n(\mathbf{p}_1, \mathbf{q}_1; \mathbf{p}_2, \mathbf{q}_2; \cdots; \mathbf{p}_n, \mathbf{q}_n; t) d\Omega_1 \, d\Omega_2 \cdots d\Omega_n =$$

$$\overline{dN(\mathbf{p}_1, \mathbf{q}_1; t) \, dN(\mathbf{p}_2, \mathbf{q}_2; t) \cdots dN(\mathbf{p}_n, \mathbf{q}_n; t)} \qquad (2.1)$$

where $d\Omega_i = d\mathbf{p}_i \, d\mathbf{q}_i$ $(i = 1, 2, \ldots, n)$ are assumed not to overlap. But wherever they do, there is a degeneracy to a lower-order density. The functions $f_1, f_2 \cdots f_n$ are known as *product* densities for the simple reason that while $f_i \, d\Omega_1 \, d\Omega_2, \ldots d\Omega_i$ denotes the probability magnitude its integral over $\Omega_1, \Omega_2, \ldots \Omega_i$ does not yield unity. It is a characteristic property of the $f_i$ that it cannot be normalized to unity in a trivial manner. The density function introduced by Boltzmann is nothing but the product density of degree one, and the usual Boltzmann equation can be obtained if we assume that

$$f_i(\mathbf{p}_1, \mathbf{q}_1; \mathbf{p}_2, \mathbf{q}_2; \cdots; \mathbf{p}_i, \mathbf{q}_i; t) = f_1(\mathbf{p}_1, \mathbf{q}_1; t) f_1(\mathbf{p}_2, \mathbf{q}_2; t) \cdots$$

$$f_1(\mathbf{p}_i, \mathbf{q}_i; t) \qquad (2.2)$$

a relation charactristic of Poisson distribution.

Next we notice that the central quantity of interest in any *macroscopic* theory is a small blob of fluid and that basic quantities like mass, momentum and local kinetic energy are defined by

$$M = \sum m \, dN(\mathbf{p}, \mathbf{q}; t)$$

$$P_i = \sum p_i \, dN(\mathbf{p}, \mathbf{q}; t) \qquad (2.3)$$

$$Q = \frac{\sum (\mathbf{p} - \mathbf{q})^2 \, dN(\mathbf{p}, \mathbf{q}; t)}{2m}$$

where the summation is over the molecules of the blob. Equations (2.3) can be interpreted in a slightly different manner. With each random point in phase-space (which corresponds to a molecule) we associate a deterministic function and the macroscopic quantities turn out to be the sum of such suitable functions. The machinery of stochastic point processes is available to us since the moments of the macroscopic quantities can be expressed as a sum of weighted integrals of the product densities. Thus the main object is to obtain the product densities of the first few orders if not some of their properties. We shall assume some kind of a coarse grain structure for the product densities and that $f_1$ and $f_2$ have been obtained by an average over the time parameter so that

$$f_1 (\mathbf{p}_1, \mathbf{q}_1;\ t) = (2\tau)^{-1} \int_{-\tau}^{\tau} \tilde{f}_1 (\mathbf{p}_1, \mathbf{q}_1;\ t + s)ds \qquad (2.4)$$

$$f_2 (\mathbf{p}_1, \mathbf{q}_1;\ \mathbf{p}_2, \mathbf{q}_2;\ t) = (2\tau)^{-1} \int_{-\tau}^{\tau} \tilde{f}_2 (\mathbf{p}_1, \mathbf{q}_1;\ \mathbf{p}_2, \mathbf{q}_2;\ t + s)ds \qquad (2.5)$$

where $\tilde{f}_1$ and $\tilde{f}_2$ are the product densities defined in the "finely" grained space. The motivation to do this is two fold: the first stems from the fact that in microscopic domain even the $S$-matrix is belived to arise from some kind of average from the fine grained domain (see, for example, Goldberger and Watson[7]) and as such it is reasonable to expect the product densities to be defined to start with, only in a coarse grained space. Secondly, it is convenient to handle the long range part of the intermolecular force whose contribution is important for fairly dense gases. Using the Markovian nature of the stochastic process, we can write down the differential equations satisfied by the product densities of the first two orders. Though the system can be viewed as a special case of a general Markovian stochastic process, the equations satisfied by the product densities will differ from the usual Chapman–Kolmogorov equation in that the transition probabilities for the system to jump from one state to another are not fixed *ab initio* but are built into the system in a self-consistent way. This is, in fact, the reason why the equations form a heirarchy. The heirarchy is broken by postulating the dependence of the product density of certain order on lower-order product densities.

We next observe that $f_1 (\mathbf{p}_1, \mathbf{q}_1;\ t)\ d\mathbf{p}_1\ d\mathbf{q}_1$ denotes the probability that a molecule has momentum between $\mathbf{p}_1$ and $\mathbf{p}_1 + d\mathbf{p}_1$ and coordinate between $\mathbf{q}_1$ and $\mathbf{q}_1 + d\mathbf{q}_1$ and $f_2 (\mathbf{p}_1, \mathbf{q}_1;\ \mathbf{p}_2, \mathbf{q}_2;\ t)d\mathbf{p}_1\ d\mathbf{q}_1\ d\mathbf{p}_2\ d\mathbf{q}_2$

denotes the joint probability that a molecule has coordinate and momentum in the range $(\mathbf{q}_1, \mathbf{q}_1 + d\mathbf{q}_1)$ and $(\mathbf{p}_1, \mathbf{p}_1 + d\mathbf{p}_1)$ and a molecule has coordinate and momentum in the range $(\mathbf{q}_2, \mathbf{q}_2 + d\mathbf{q}_2)$ and $(\mathbf{p}_2, \mathbf{p}_2 + d\mathbf{p}_2)$. Using the property that $f_1(\mathbf{p}_1, \mathbf{q}_1; t)d\mathbf{p}_1 \, d\mathbf{q}_1$ is a probability magnitude, we find that $f_1$ in the absence of external forces satisfies the equation

$$\frac{\partial f_1}{\partial t} + \left(\frac{p_{1t}}{m}\right)\frac{\partial f_1}{\partial q_{1t}} = J_c + J_s \tag{2.6}$$

where $J_c$ denotes the contribution from collisions and is given by

$$J_c = \iint [f_2 \, (\mathbf{p}_1 - \Delta\mathbf{p}_1, \, \mathbf{q}_1; \, \mathbf{p}_2, \, \mathbf{q}_1 + \sigma\mathbf{k}; \, t)$$

$$- f_2 \, (\mathbf{p}_1, \, \mathbf{q}_1; \, \mathbf{p}_2, \, \mathbf{q}_1 - \sigma\mathbf{k}; \, t)]v_{12}b \, db \, d\epsilon \, d\mathbf{p}_2 \tag{2.7}$$

where $\sigma$ is the diameter of a molecule and $\mathbf{k}$ is the unit vector along the direction of the line of centers. $J_s$ is the contribution from a force term which is due to the Brownian-like motion of the molecule due to other molecules over a relaxation time $\tau \ll \Delta t$ and is given by

$$J_s = -(2\tau)^{-1} \int_{-\tau}^{\tau} F_i^{12} \frac{\partial \tilde{f}_2}{\partial p_{1t}} \, (\mathbf{p}_1, \, \mathbf{q}_1; \, \mathbf{p}_2, \, \mathbf{q}_2; \, t + s)ds \tag{2.8}$$

where $F^{12}$ is the intermoleculer force. For a gas which is not too dense (2.8) gives negligible contribution and (2.7) is the only appreciable contribution so that (2.6) becomes the usual Boltzmann equation with collision term if we make the Poisson approximation

$$f_2 \, (\mathbf{p}_1, \, \mathbf{q}_1; \, \mathbf{p}_2, \, \mathbf{q}_2; \, t) = f_1 \, (\mathbf{p}_1, \, \mathbf{q}_1; \, t) \, f_1 \, (\mathbf{p}_2, \, \mathbf{q}_2; \, t) \tag{2.9}$$

Equations similar to (2.6) were first noticed by Kirkwood[8] and Born and Green.[9] However, they dealt with the distribution function for fixed number of molecules where (2.3) is valid for product densities. The terms $J_c$ and $J_s$ are of course defined in terms of the product density of degree of two without explicit reference to the number of molecules in the system. Thus (2.6) is valid for all types of fluids.

In an exactly similar manner, we can write down the equation satisfied by $f_2$. However, in this case a molecule with position in the cell $(\mathbf{q}_1, \mathbf{q}_1 + d\mathbf{q}_1)$ experiences a force $F^{12}$ due to another molecule in the cell $(\mathbf{q}_2, \mathbf{q}_2 + d\mathbf{q}_2)$ and vice versa. Moreover, $J_c$ and $J_s$ will involve product densities of third order. Thus $f_2$ satisfies the equation

$$\frac{\partial f_2}{\partial t} + \frac{p_{1t}}{m}\frac{\partial f_2}{\partial q_{1t}} + \frac{p_{2t}}{m}\frac{\partial f_2}{\partial q_{2t}} + F_i^{12}\frac{\partial f_2}{\partial p_{1t}} + F_i^{21}\frac{\partial f_2}{\partial p_{2t}} = J_c + J_s \tag{2.10}$$

where

$$J_c = \iint [f_3(\mathbf{p}_1 - \Delta\mathbf{p}_1, \mathbf{q}_1 - \Delta\mathbf{q}_1; \mathbf{p}_2 - \Delta\mathbf{p}_2, \mathbf{q}_2 - \Delta\mathbf{q}_2; \mathbf{p}_3 - \Delta\mathbf{p}_3, \mathbf{q}_1 + \sigma\mathbf{k}; t)$$

$$- f_3 (\mathbf{p}_1, \mathbf{q}_1; \mathbf{p}_2, \mathbf{q}_2; \mathbf{p}_3, \mathbf{q}_1 - \sigma\mathbf{k}; t)]b \, db \, d\epsilon \, v_{13} \, d\mathbf{p}_3 \qquad (2.11)$$

$$J_s = -(2\tau)^{-1}\int_{-\tau}^{\tau} \iint F_i^{13(s)} \frac{\partial}{\partial p_{1i}} \bar{f}_3 (\mathbf{p}_1, \mathbf{q}_1; \mathbf{p}_2, \mathbf{q}_2; \mathbf{p}_3, \mathbf{q}_3; t + s) \, d\mathbf{p}_3 \, d\mathbf{q}_3 \, ds$$

$$- (2\tau)^{-1}\int_{-\tau}^{\tau} \iint F_i^{24(s)} \frac{\partial}{\partial p_{2i}} \bar{f}_3 (\mathbf{p}_1, \mathbf{q}_1; \mathbf{p}_2, \mathbf{q}_2; \mathbf{p}_4, \mathbf{q}_4; t + s) \, d\mathbf{p}_4 \, d\mathbf{q}_4 \, ds$$

$$(2.12)$$

We can proceed in this manner and obtain an infinite set of coupled equations for the product densities which appear in a form similar to the *B-B-G-K-Y* heirarchy. *However, the functions defined above have a different interpretation and the moments of physical quantities of interest are appropriate weighted integrals of these product densities* as has been explained in Ref. 2. But since the main characteristic features of the equations like nonlinearity or coupling to higher order densities is exactly the same as those enjoyed by *B-B-G-K-Y* heirarchy, it is indeed difficult to obtain nonperturbative solution valid for configurations fairly far from equilibrium position even if we make suitable assumptions regarding the functional dependence of $f_3$ and $f_2$ on the first order density. In view of this inherent difficulty, we attempt to study the onset of instability by concentrating our attention on small deviations from the equilibrium configuration. We hope this will throw some light on the configurations which are appreciably different from equilibrium. In the next section, we shall deal with dense fluids and indicate the onset of instability. Guided by the main results of dense fluid theory, we will formulate in Section 5 a theory of turbulence on the basis of the stirring force and random response coefficients.

## 3. DENSE FLUID APPROXIMATION: ONSET OF INSTABILITY

We proceed to a solution of equation (2.6) under certain approximations. The first one consists in closing the heirarchy at the first stage itself. $f_2$ is approximated by

$$f_2(\mathbf{p}_1, \mathbf{q}_1; \mathbf{p}_2, \mathbf{q}_2; t) = g(\mathbf{q}_1, \mathbf{q}_2)f_1(\mathbf{p}_1, \mathbf{q}_1; t)f_1(\mathbf{p}_2, \mathbf{q}_2; t) \qquad (3.1)$$

where $g(\mathbf{q}_1, \mathbf{q}_2)$ is the pair correlation function and corresponds to the factor $\chi$ introduced in Ref. 2. Under this approximation, we estimate the contributions from the collision term. The calculations have been done by Green[10] explicitly in connection with the theory of liquids for the Born–Green functions. In fact Rice and his collaborators[11,12] have made explicit use of these results to obtain refined transport coefficients for liquid Ar and other inert gases. Using the techniques employed in Refs. 10 and 12 we find

$$J_c = J_1 + J_2 \tag{3.2}$$

$$J_1 = g_0(\mathbf{q}_1, \sigma) \iint [f_1(\mathbf{p}_1 - \Delta\mathbf{p}_1, \mathbf{q}_1; t) f_1(\mathbf{p}_2 - \Delta\mathbf{p}_2, \mathbf{q}_1; t)$$

$$- f_1(\mathbf{p}_1, \mathbf{q}_1; t) f_1(\mathbf{p}_2, \mathbf{p}_2; t)] v_{12} b \, db \, d\epsilon \, d\mathbf{p}_2$$

$$J_2 = g_0(\mathbf{q}_1, \sigma) \iint \left[ f_1(\mathbf{q}_1 - \Delta\mathbf{p}_1, \mathbf{q}_1; t)\sigma k_i \frac{\partial f_1(\mathbf{p}_2 - \Delta\mathbf{p}_2, \mathbf{q}_1; t)}{\partial q_{1i}} \right.$$

$$\left. + f_1(\mathbf{p}_1, \mathbf{q}_1; t)\sigma k_i \frac{\partial f_1(\mathbf{p}_2, \mathbf{q}_1; t)}{\partial q_{1i}} \right] v_{12} b \, db \, d\epsilon \, d\mathbf{p}_2 \tag{3.3}$$

where $g_0$ is the equilibrium pair correlation function, the other symbols in (3.3) being defined as in Ref. 10. In the estimation of the terms arising from the soft force, we again use the method of Green and Kirkwood. After some calculation, we find

$$J_s = -F_{1i}^* \frac{\partial f_1(\mathbf{p}_1, \mathbf{q}_1; t)}{\partial p_{1i}} + \zeta a f_1(\mathbf{p}_1, \mathbf{q}_1; t) \tag{3.4}$$

where

$$\mathbf{F}_1^* = (2\tau)^{-1} \int_{-\tau}^{\tau} \int g_2(\mathbf{q}_1, \mathbf{q}_2) \vec{f}_1(\mathbf{p}_2, \mathbf{q}_2; t+s) F^{12}(q_{12}) d\mathbf{q}_2 \, d\mathbf{p}_2$$

$$= \langle F_1^{(s)} \rangle + F_1^{(d)} \tag{3.5}$$

$$\zeta = (6kT\tau)^{-1} \int_{-\tau}^{\tau} ds \int_{-s}^{0} ds' \iint F_1^{12}(q_{12}) F_i^{(s)} \left( q + \overline{s+s'} \frac{\mathbf{p}_2}{m} \right)$$

$$g_2(\mathbf{q}_1, \mathbf{q}_2) f_1(\mathbf{p}_2, \mathbf{q}_2; t) d\mathbf{p}_2 \, d\mathbf{q}_2 \tag{3.6}$$

$$a = \frac{\partial}{\partial p_{1i}} \left\{ \left( \frac{p_{1i}}{m} - u_i \right) + kT \frac{\partial}{\partial p_{1i}} \right\} \tag{3.7}$$

where $u$ is the average velocity of a macroscopic blob of fluid and $T$ is the local temperature in equilibrium state. We have followed closely the notation of Rice and Alnatt.[11] $\langle F_1^{(s)} \rangle$ is the net soft force experienced by a molecule and is evaluated through the use of equilibrium pair correlation function $F_1^{(d)}$ standing for the contribution due to deviation of $g(q_1, q_2)$ from $g_0(q_1, q_2)$. The complicated nature of $J_s$ is essentially due to the occurrence of the fine grained product density $f_1$ rather than $\bar{f}_1$ in the integrand. The temporal and spatial evolution of $\bar{f}_2$ during the relaxation time has to be estimated as demonstrated by Green[10] (see also Rice and Alnatt[11]) and this leads us to the identification of various terms as given by equations (3.4) to (3.7). Collecting all the terms, we can rewrite (2.3) as

$$\frac{\partial f_1}{\partial t} + \frac{p_{1t}}{m}\frac{\partial f_1}{\partial q_{1t}} + F_{1t}^*\frac{\partial f_1}{\partial p_{1t}} = \zeta\frac{\partial}{\partial p_{1t}}\left\{\left(\frac{p_{1t}}{m} - u_t\right) + kT\frac{\partial}{\partial p_{1t}}\right\}f_1 + J_1 + J_2$$

(3.8)

which is identical with the equation obtained by Rice and Alnatt, the difference arising from the interpretation of the function $f_1$. Equation (3.8) can be solved only by a perturbation technique and as such the results are valid for small deviations from Maxwellian distribution. This has been achieved in Ref. 12 and the explicit solution is given by

$$f_1(\mathbf{p}, \mathbf{q}; t) = f_1^0(\mathbf{p}, \mathbf{q}; t)\left\{1 + A\left(\frac{5}{2} - W^2\right)W_t\frac{\partial lnT}{\partial q_t}\right.$$

$$\left. + B\left(W_t W_j - \frac{1}{3}W^2\delta_{ij}\right)\frac{1}{2}\left(\frac{\partial u_i}{\partial q_j} + \frac{\partial u_j}{\partial q_i}\right)\right\}$$

(3.9)

where

$$A = \left(\frac{15}{4}\right)\frac{\{1/g(\sigma) + 2\pi n\,\sigma^3/5\}}{\{4n\Omega^{(2,2)} + (45/4)[\varphi/mg(\sigma)]\}}$$

(3.10)

$$B = 5\frac{\{1/g(\sigma) + 4\pi n\sigma^3/5\}}{\{4n\Omega^{(2,2)} + 5\varphi/mg(\sigma)\}}$$

(3.11)

$$\mathbf{W} = \left(\frac{m}{2kT}\right)^{1/2}\left(\frac{\mathbf{p}}{m} - \mathbf{u}\right)$$

(3.12)

and $g(\sigma)$ is the pair correlation function evaluated on the sphere with radius equal to the molecular diameter. $\Omega^{(2,2)}$ is the reduced collision cross section and is given by

$$\Omega^{(2,2)} = \left(\frac{4\pi kT}{m}\right)^{1/2}\iint\gamma^7(1 - \cos^2\chi)e^{-\gamma^2}b\,db\,d\gamma$$

(3.13)

where

$$\gamma = \left(\frac{m}{4\pi kT}\right)^{1/2}v_{12}$$

(3.14)

and $\chi$ is the angle of deflection in a collision.

The contributions to the pressure tensor and the heat flux vector arising from the perturbation as given by (3.9) have been discussed in great detail by Chapman and Cowling.[13] We, however, would like to confine ourselves to the development of the fluctuation of the macroscopic variables like mass, momentum, pressure, and temperature about their mean values. At the outset, we wish to observe that (3.1) completely determines the character of the fluctuation of physically measurable quantities. To illustrate this point, let us consider $M$ the mass of a macroscopic elemental blob of fluid centerd at $r$. The mean square value of $M$ is given by

$$\bar{M}^2 = \iint m^2 f_1(\mathbf{p}_1, \mathbf{q}_1; t) d\mathbf{p}_1 \, d\mathbf{q}_1$$

$$+ \iiiint m^2 f_2(\mathbf{p}_1, \mathbf{q}_1; \mathbf{p}_2, \mathbf{q}_2; t) d\mathbf{p}_1 \, d\mathbf{q}_1 \, d\mathbf{p}_2 \, d\mathbf{q}_2 \qquad (3.15)$$

where the integration over $q$ is to be performed over a sphere centred at $r$ and radius $dr$ defined with macroscopic accuracy. Using (3.1) we find

$$\bar{M}^2 = m\bar{M} + \left(\frac{\bar{M}}{\Omega}\right)^2 \iint g(\mathbf{q}_1, \mathbf{q}_2) d\mathbf{q}_1 \, d\mathbf{q}_2 \qquad (3.16)$$

where $\Omega$ is a macroscopically elemental volume in space. For a dilute gas, $g$ is always equal to unity and hence

$$\bar{M}^2 = (\bar{M})^2 + m\bar{M} \qquad (3.17)$$

which leads to the fluctuation theorem of statistical mechanics. If on the other hand, $g$ differs from unity even by a few percent, this contributes *an equal fraction of mean square mass to the fluctuation.* In fact as we pass on from a dilute gas to a dense fluid, $g$ increases with increasing density (see, for example, Chapman and Cowling[13]). In such a case the collision term $J_c$ on the right-hand side of (3.8) does not give appreciable contribution since for such a dense fluid the molecules are so closely packed together that collisions are rare, the dominant contributions to the right-hand side arising from $J_s$ only. Apart from the magnitude of $g$ which increases beyond limit, *its range which is normally of the order of a few molecular diameters increases well beyond its microscopic limits* (see, for example, Munster[14]). Although this point has been emphasized in proper perspective in the Born–Green–Kirkwood theory of fluids, *the overall effect on the second-order correlation function and its consequent impact on the fluctuations of macroscopic variables has been* completely lost sight of in the innumerable attempts to arrive at the higher moments of the function $f_1(\mathbf{p}_1, \mathbf{q}_1; t)$. On the contrary, even a complete knowledge of all the moments of $f_1(\mathbf{p}_1, \mathbf{q}_1; t)$ does not throw as much light as does the limited information contained in an assumption like (3.1). In the next section, we shall study the effect of (3.1) on momentum and pressure correlations.

To get an estimate of the fluctuation, we use the first approximation given by Chapman and Cowling[13] and replace $g(\mathbf{q}_1, \mathbf{q}_2)$ by

$$\chi = 1 + 0.625\rho b + 0.2869\rho^2 b^2$$

$$\rho b = \frac{2\pi n\sigma^3}{3} \tag{3.18}$$

where $n$ is the average number density of molecules. Using (3.18) in (3.16), we find

$$\frac{\bar{M}^2 - (\bar{M})^2}{(\bar{M})^2} = \rho b(0.625 + 0.2869\rho b) \tag{3.19}$$

Using the density viscosity data for carbon dioxide† as given by Chapman and Cowling (see Table 29), we find the right-hand side of (3.19) has values 0.019, 0.073, 0.129, and 0.282 corresponding to pressures 15.37, 57.6, 104.5, and 212.4 times the atmospheric pressure. Thus the mean square deviation of mass density is of the same order as the mean square value. These results are equally applicable to other macroscopic quantities like absolute value of the momentum, local energy, and pressure. This clearly shows that instability sets in. The next question we would like to ask is: How does the instability make itself manifest? This can be answered in a positive manner when the gas is in motion. Let us assume that the macroscopic velocity is of the order of $10^4$ cm/sec. Then we can calculate the Reynolds number for the gas in the state of motion and find that it is of the order of $10^7$ or $10^8$, and it is a well-known fact that the motion of the gas is turbulent for the Reynolds number in that range. *Thus the instability in this case manifests itself in the form of turbulent motion.* In fact, as $\chi$ increases and becomes unbounded, the motion of the fluid is in a highly developed state of turbulence.

---

†In this case the temperature is in the neighborhood of the critical temperature, the long-range nature of a correlation leading to a nonvanishing value of $g - 1$ over a considerable part of the phase space.

It is interesting to observe that *by progressively increasing $\chi$, or in other words by progressively increasing the Reynolds number, we can attain a turbulent flow starting from a laminar flow.* If on the other hand, the fluid is not in a state of motion, the instability must make itself manifest by a change of state.[14] Thus the gas must condense into a liquid, the transition being attained in a gradual manner as $\chi$ is gradually increased. The arguments are only qualitative and we hope to amplify the same by quantitative estimates. In this particular case, we have to bring to the fore the instability in the absolute value of pressure and temperature, and these in turn by the very nature of the dense fluid approximation are functionals of the higher-order product densities, since there is a substantial contribution from the potential force. The explicit dependence of these macroscopic quantities on the higher-order product densities will be brought out in the next section when we deal with the methods of calculation of hydrodynamical correlations. We shall also show how the general properties of higher-order product densities enable us to distinguish between different manifestations of the instability. In particular we shall be able to distinguish between phase transition and onset of turbulent motion.

## 4. SECOND-ORDER APPROXIMATION TO A DENSE FLUID

The discussion in Section 3 had been based on the equation (3.1) which explains the functional dependence of the second-order product density on the first order density. One of the main conclusions arising from (3.1) is that the fluctuation about the mean mass density is of the same order of magnitude as that of the square of the mean. Though this is a useful result, this does not completely determine the distribution structure. In fact it is well known that in turbulent flows behind a grid the flatness factor and skewness show a distinct departure from the Gaussian value even though the mean square deviation of the absolute value of velocity is Gaussian (see, for example, Lin[15]). Hence it is necessary to have some idea of the behavior of higher-order product densities. There is also another good reason for the introduction of higher-order densities. As the density of the gas is increased, the potential energy becomes an important criterion and macroscopic quantities like pressure and local energy are strongly dependent on the same. In Section 2, we defined the macroscopic variables as stochastic integrals involving the random variable $dN(\mathbf{p}, \mathbf{q}, t)$ which represents the number of molecules in the elemental phase space around $(\mathbf{p}, \mathbf{q})$. For a dense gas it is necessary to modify the definition. The following equations are self-explanatory:

$$\mathscr{P}_{ij} = \sum m W_{1i} W_{1j} dN(\mathbf{p}_1, \mathbf{q}_1; t)$$

$$+ \frac{1}{2} \sum \int m F_i^{12}(\mathbf{q}_{12}) q_{12j} \, dN(\mathbf{p}_1, \mathbf{q}_1; t) dN(\mathbf{p}_2, \mathbf{q}_2; t) \qquad (4.1)$$

$$Q = \frac{1}{2} \sum m W_1^2 \, dN(_1\mathbf{p}, \mathbf{q}_1; t)$$

$$+ \frac{1}{2} \sum \int V_{12}(\mathbf{q}_{12}) dN(\mathbf{p}_1, \mathbf{q}_1; t) dN(\mathbf{p}_2, \mathbf{q}_2; t) \qquad (4.2)$$

where $\mathbf{q}_{12} = \mathbf{q}_1 - \mathbf{q}_2$ and the summation is taken over the molecules of the blob and the integration over the phase space of a molecule with coordinates $(\mathbf{p}_2, \mathbf{q}_2)$. $\mathbf{F}^{12}(\mathbf{q}_{12})$ is the intermolecular force and $V_{12}(\mathbf{q}_{12})$ is the potential due to two molecules at positions $\mathbf{q}_1$ and $\mathbf{q}_2$. The second term on the right-hand side of (4.1) and (4.2) is non-zero only when the elemental phase space of the two molecules do not overlap. Using this fact, we find

$$\bar{\mathscr{P}}_{ij} = \int m W_{1i} W_{1j} f_1(\mathbf{p}_1, \mathbf{q}_1; t) dp_1 dq_1$$

$$+ \frac{1}{2} \int m F_i^{12} q_{12j} f_2(\mathbf{p}_1, \mathbf{q}_1; \mathbf{p}_2, \mathbf{q}_2; t) dp_1 \, dq_1 \, dp_2 \, dq_2 \qquad (4.3)$$

$$\bar{Q} = \frac{1}{2} \int m W_1^2 f_1(\mathbf{p}_1, \mathbf{q}_1; t) dp_1 \, dq_1$$

$$+ \frac{1}{2} \int V_{12}(\mathbf{q}_{12}) f_2(\mathbf{p}_1, \mathbf{q}_1; \mathbf{p}_2, \mathbf{q}_2; t) dp_1 \, dq_1 \, dp_2 \, dq_2 \qquad (4.4)$$

where the domain of integration over $\mathbf{q}$'s is defined as in (3.15). The above equations bring out the fact that the mean values of $\mathscr{P}_{ij}$ and $Q$ are functionals of the product densities of the first two orders. In fact, in the liquid state the dominant contribution comes only from the term involving second-order product density. It is clear that the mean square is a functional of the densities of the first four orders. Thus the fluctuations of $\mathscr{P}_{ij}$ and $Q$ can be determined only if we make suitable assumptions on the functional dependence of the fourth-order density on the lower-order densities.

From the above considerations, it follows that the approximation (3.1) even if it is valid may not be sufficient to determine the fluctuation in local energy. Since we need at least up to the fourth-order density, it is necessary to set up equations for $f_3$ and close the heirarchy at that stage by the approximation

$$f_4(\mathbf{p}_1, \mathbf{q}_1; \mathbf{p}_2, \mathbf{q}_2; \mathbf{p}_3, \mathbf{q}_3; \mathbf{p}_4, \mathbf{q}_4; t)$$

$$= \sum_{1234} \phi(\mathbf{q}_1, \mathbf{q}_2, \mathbf{q}_3, \mathbf{q}_4) f_2(\mathbf{p}_1, \mathbf{q}_1; \mathbf{p}_2, \mathbf{q}_2; t) f_2(\mathbf{p}_3, \mathbf{q}_3; \mathbf{p}_4, \mathbf{q}_4; t) \qquad (4.5)$$

where some general assumptions regarding $\phi$ can be made consistent with the magnitude of the fluctuations that are to be expected. Equation (4.5) is the least that can be done in such a situation and again other statistical quantities like skewness and flatness are not determined within the approximation scheme. However it should not be imagined that the scheme of calculations envisaged is fairly simple or even possible for configurations which are sufficiently far from equilibrium position. This point can be illustrated by confining our attention to the Born–Green theory as formulated through Mayer's clusters. In this case, we try to solve (2.7) by assuming

$$f_3(\mathbf{p}_1, \mathbf{q}_1; \mathbf{p}_2, \mathbf{q}_2; \mathbf{p}_3, \mathbf{q}_3; t)$$

$$= \frac{f_2(\mathbf{p}_1, \mathbf{q}_1; \mathbf{p}_2, \mathbf{q}_2; t) f_2(\mathbf{p}_2, \mathbf{q}_2; \mathbf{p}_3, \mathbf{q}_3; t) f_2(\mathbf{p}_1, \mathbf{q}_1; \mathbf{p}_3, \mathbf{q}_3; t)}{f_1(\mathbf{p}_1, \mathbf{q}_1; t) f_1(\mathbf{p}_2, \mathbf{q}_2; t) f_1(\mathbf{p}_3, \mathbf{q}_3; t)} \qquad (4.6)$$

in addition to (3.1). The calculations have been done by Rice and Alnatt[11] who have obtained explicit solution for the second-order Born–Green distribution functions. Since equations (2.7) to (2.9) are very similar to those of Rice and Alnatt, we can take over their results and adapt it to the present situation. Under the approximation (4.6) and (2.3), (2.7) reduces to

$$\left( \frac{\partial}{\partial t} + \frac{p_{1i}}{m} \frac{\partial}{\partial q_{1i}} + \frac{p_{2i}}{m} \frac{\partial}{\partial q_{2i}} + F_{1i}^{(2)} \frac{\partial}{\partial p_{1i}} + F_{2i}^{(2)} \frac{\partial}{\partial p_{2i}} \right) f_2(\mathbf{p}_1, \mathbf{q}_1; \mathbf{p}_2, \mathbf{q}_2; t)$$

$$= J_1 + J_2 + \{ \zeta(\mathbf{q}_1) a_1 + \zeta(\mathbf{q}_2) a_2 \} f_2(\mathbf{p}_1, \mathbf{q}_1; \mathbf{p}_2, \mathbf{q}_2; t) \qquad (4.7)$$

where $a_j$ is given by

$$a_j = \frac{\partial}{\partial p_{ji}} \left\{ \left( \frac{p_{ji}}{m} - u_i(q_j) \right) + kT(q_j) \frac{\partial}{\partial p_{ji}} \right\} \qquad (4.8)$$

and $\zeta$, $J_1$ and $J_2$ are as defined earlier (Section 3). The solution for small deviations from equilibrium position is given by

$$f_2(\mathbf{p}_1, \mathbf{q}_1; \mathbf{p}_2, \mathbf{q}_2; t) = g(\mathbf{q}_1, \mathbf{q}_2) f_1(\mathbf{p}_1, \mathbf{q}_1; t) f_1(\mathbf{p}_2, \mathbf{q}_2; t)$$

$$\left\{ 1 - \sum_{j=1}^{2} \left[ \alpha_j \left( \frac{5}{2} - W_j^2 \right) W_{ji} \frac{\partial}{\partial q_{ji}} \ln T(\mathbf{q}_j) \right. \right.$$

$$+ \beta_j \left( W_{jk} W_{ji} - \frac{1}{3} W_j^2 \delta_{kl} \right) \frac{1}{2} \left( \frac{\partial}{\partial q_{jk}} u_i(\mathbf{q}_j) \right.$$

$$\left. \left. + \frac{\partial}{\partial q_{ji}} u_k(\mathbf{q}_j) \right) + C_j W_{ji} G_{ji} \right] \right\} \qquad (4.9)$$

where $\alpha_j$ and $\beta_j$ are certain coefficients similar to $A$, $B$ defined in (3.10) and (3.11). Now, $C_j$ is given by

$$C_j = -\frac{(2m/kT(\mathbf{q}_j))^{1/2}}{\zeta(\mathbf{q}_j)} \tag{4.10}$$

and $\mathbf{G}_j$ is the fluctuation of the force experienced by a molecule about $\mathbf{F}^*$ which, in turn, is defined by (3.5).

At the outset, we notice that though the solution (4.9) is perturbative, a number of useful inferences can be drawn. For instance, the force term $\mathbf{G}_j$ gives a nonvanishing contribution to fluctuations in momentum showing thereby the importance of higher order approximations. Likewise there is a nonvanishing contribution to the heat flux and therefore to the transport coefficients. Of course, the first term which contains the pair correlation function as a factor yields the important contribution that explains the build up of the fluctuation.

The function $\phi$ defined through (4.5) plays an important role in that its behavior characterizes the type of instability the system is subject to. If the magnitude of $\phi$ is of order unity in a substantial part of the phase space, we can show that the pressure and local energy (temperature) do not fluctuate about their corresponding mean values so that their mean values are their actual values. This can be readily seen by observing that the mean square value of local energy is given by

$$\overline{Q^2(r)} = \text{contribution from a weighted integral of first-order}$$
$$\text{product density} +$$

$$+ \frac{1}{4} \iint m^2 W_1^2 W_2^2 f_2(\mathbf{p}_1, \mathbf{q}_1 ; \mathbf{p}_2, \mathbf{q}_2 ; t)\, d\mathbf{p}_1\, d\mathbf{q}_1\, d\mathbf{p}_2\, d\mathbf{q}_2$$

$$+ \frac{1}{4} \iint [V_{12}(\mathbf{q}_{12})]^2 f_2(\mathbf{p}_1, \mathbf{q}_1 ; \mathbf{p}_2, \mathbf{q}_2 ; t)\, d\mathbf{p}_1\, d\mathbf{q}_1\, d\mathbf{p}_2\, d\mathbf{q}_2$$

$$+ \frac{1}{4} \iiint m W_3^2 V_{12}(\mathbf{q}_{12}) f_3(\mathbf{p}_1, \mathbf{q}_1 ; \mathbf{p}_2, \mathbf{q}_2 ; \mathbf{p}_3, \mathbf{q}_3 ; t)$$

$$d\mathbf{p}_1\, d\mathbf{q}_1\, d\mathbf{p}_2\, d\mathbf{q}_2\, d\mathbf{p}_3\, d\mathbf{q}_3$$

$$+ \frac{1}{4} \iiiint V_{12}(\mathbf{q}_{12}) V_{34}(\mathbf{q}_{34})$$

$$f_4(\mathbf{p}_1, \mathbf{q}_1 ; \mathbf{p}_2, \mathbf{q}_2 ; \mathbf{p}_3, \mathbf{q}_3 ; \mathbf{p}_4, \mathbf{q}_4 ; t)$$

$$d\mathbf{p}_1\, d\mathbf{q}_1\, d\mathbf{p}_2\, d\mathbf{q}_2\, d\mathbf{p}_3\, d\mathbf{q}_3\, d\mathbf{p}_4\, d\mathbf{q}_4 \tag{4.11}$$

Since the dominant contribution to the mean square value comes from the last term of equation (4.11) it follows from the result that $\phi \approx 1$ that the mean square value of $Q$ is nothing but the square of the mean. In an exactly similar manner we can show that the pressure $P_{ij}$ does not fluctuate about its mean value. Thus, if the fluid is at rest and if the value assumed by and the range of $g(\mathbf{q}_1, \mathbf{q}_2)$ increases the system will exhibit density fluctuations, the pressure and temperature remaining stable. This is precisely the manner in which a gas condenses. If it were possible to prove that under these conditions, the mean value of pressure (which is also its true value) increases with $g$ and the mean value of temperature is in a neighborhood of the critical temperature, then we would have reproduced theoretically the characteristics of phase transition. Since these in turn depend on the weighted integrals of the product densities of the first few orders, a proof of the unbounded nature of $g$ on the basis of generalized Boltzmann equations, for the product densities would constitute a sound kinetic basis for the theory of phase transition.

If however, on the other hand, in addition to $g$, $\phi$ also increases beyond unity in an appreciable part of the phase space, both the pressure and local energy fluctuates appreciably about their corresponding mean values. Such a situation corresponds to turbulent motion, the instability in this case characterizing a transition in the state of motion of the fluid. It may be worthwhile to extend the analogy of phase transition and examine whether there are some macroscopic dynamical quantities that are stable. This is a very difficult question since the macroscopic quantities that can be obtained from a microscopic theory are very limited indeed. Finally, we wish to observe that the technique we have developed in this section can be readily used to incorporate three-body forces if they were significant. This can be achieved by adding an appropriately weighted stochastic integral involving products of three $dN(p, q ; t)$ to (4.1) and (4.2).

## 5. DEVELOPMENT OF TURBULENT MOTION

As pointed out in Ref. 2, it is convenient to start with the stochastic integrals for momentum, mass, and local energy rather than the usual hydrodynamical equations satisfied by the mean values of these quantities. In fact, we can go a step further and remark that *the hydrodynamical equations are not characteristic of the type of motion suffered by the fluid unless the fluctuation theorem of statistical mechanical systems is shown to be valid.* Using the solution (4.9) and definition of momentum as a stochastic integral as given in Section 2, we find that the second-order correlation in momentum is given by

$$\overline{P_i(r_1)P_j(r_2)} = \text{contribution from the first-order product density}$$

$$+ \ \overline{P_i(r_1)} \ \overline{P_j(r_2)} \iint \frac{g(q_1, q_2)}{\Omega^2} dq_1 \, dq_2$$

$$+ \ \frac{\overline{P_j(r_2)}}{\zeta(r_1)} \int \frac{mn(q_1)}{\Omega} G_{1i}(q_1)g_{r_2}(q_1)dq_1$$

$$+ \ \frac{\overline{P_j(r_1)}}{\zeta(r_2)} \int \frac{mn(q_2)}{\Omega} G_{1j}(q_2)g_{r_1}(q_2)dq_2 \tag{5.1}$$

where $g_{r_1}(q_2)$ is the value of the pair correlation function when one variable $q_1$ is integrated over the sphere with center at $r_1$.

In obtaining (5.1) we have used the perturbation solution of the second-order product density and as such (5.1) cannot be expected to convey much information for a fairly developed state of turbulence. On the other hand, we can make useful inferences by looking at the deviations from laminar flow properties. The first term on the right-hand side arises from the overlap of the phase space occupied by two molecules and can be estimated as per the formula given in Ref. 16. However, this is smaller by an order of magnitude equal to the number of molcules contained in a macroscopically elemental volume and as such we have not paid any attention to it. The second term clearly shows the origin of the correlation, and if we make a crude estimate by using (3.18), we then find that the correlation is of the order of a few percent, at least to start with, and can be expected to rise further when the motion becomes sufficiently unstable. This term *readily brings out the microscopic origin of turbulence.* The last two terms are a little bit difficult to estimate exactly without any assumption regarding the intermolecular forces and the consequent Brownian motion to which the molecules are subjected. However, their origin can be traced to the non-zero fluctuation of the force experienced by a molecule in the vicinity of another molecule.

Equation (5.1) does not throw much light on the momentum (velocity) correlation if $r_1$ and $r_2$ are separated by a distance of macroscopic order. In fact, there will apparently be no correlation for such distances and $\overline{P_i(\mathbf{r}_1)\,P_j(\mathbf{r}_2)}$ appear to factor out as $\overline{P_i(\mathbf{r}_1)}\;\overline{P_j(\mathbf{r}_2)}$. It should not however be imagined that the correlations are purely microscopic in nature with no consequences on macroscopic velocities. This point can be made clear by considering the momentum correlation at two points which are separated by a distance of infinitesimal order on a macroscopic scale. Then there will be an overlap of the volume elements at $\mathbf{r}_1$ and $\mathbf{r}_2$, and the momentum of the blob of fluid centerd at $\mathbf{r}_1$ is correlated to the momentum of the blob of fluid centerd at $\mathbf{r}_2$. Such a correlation should not be construed as purely microscopic in nature and, in fact, *by moving the blob in this manner we obtain correlations for macroscopic velocities between two points separated by a sizable macroscopic distance.* It should be borne in mind that *such a correlation is essentially due to the nonequilibrium configuration of the blob of fluid and in fact if the system is in local equilibrium no such correlation between two blobs of fluid persists even if they were to be separated by an infinitesimal distance* (in the macroscopic scale).

Next it is worthwhile to seek the macroscopic consequences of the breakdown of local equilibrium when $g(\mathbf{q}_1, \mathbf{q}_2)$ and $\phi(\mathbf{q}_1, \mathbf{q}_2, \mathbf{q}_3, \mathbf{q}_4)$ differ from unity over a substantial region of the phase-space. One such property is the existence of the correlation between momentum of the blobs at two different directions at the same macroscopic point. This can be obtained from (5.1) by setting $\mathbf{r}_1 = \mathbf{r}_2$. Such correlations do exist in the case of nonisotropic turbulence and can be measured experimentally. The second term of the right-hand side of (5.1) yields the dominant contribution to the correlation and, hence, an experimental determination of the velocity correlation will help us determine the pair correlation function. Once the pair correlation function is obtained, we can derive explicit expressions for other macroscopic variables like stress correlations corresponding to different pairs of directions at the same macroscopic point. We can also make definite predictions regarding the fluctuation of local energy (temperature). A detailed study of these quantities will also enable us to set an idea of the distribution of the force $\mathbf{F}^*$ experienced by a molecule in the vicinity of another molecule. However it is difficult to make any definite statement on its role in the turbulent region.

Next we wish to seek an alternative method of describing the dynamics of turbulent flows since the approach through the stochastic integrals has turned out to be unwieldy due to our inability to solve for the product densities corresponding to configurations far from equilibrium position. Using the definition of average, mass, momentum, and local energy explained in Section 3, we readily obtain from (3.8) the following equations of motion depicting the average behavior of a blob of fluid:†

$$\frac{\partial \bar{M}}{\partial t} + \frac{\partial \bar{P}_i}{\partial q_i} = 0 \tag{5.2}$$

$$\frac{\partial \bar{P}_i}{\partial t} + \frac{\partial}{\partial q_j}\left(\frac{\bar{M}\,\bar{P}_i\,\bar{P}_j}{m^2}\right) = -\frac{\partial}{\partial q_j}\bar{\mathscr{P}}_{ij} + F_i^* + K_i \tag{5.3}$$

$$\frac{\partial \bar{Q}}{\partial t} + \frac{\partial}{\partial q_i}\left(\frac{\bar{P}_i\,\bar{Q}}{\bar{M}}\right) = -\frac{\partial \bar{J}_i}{\partial q_i} - \bar{\mathscr{P}}_{ij}\frac{\partial}{\partial q_j}\left(\frac{\bar{P}_i}{\bar{M}}\right)$$

$$+ 2(u_i(\mathbf{q})\bar{F}_i^* - \overline{F_i^* v_i}) + L \tag{5.4}$$

$\bar{P}_{ij}$ is evaluated to first order in the dilute gas approximation:

$$\bar{P}_{ij} = \frac{2}{3}\,\bar{Q}\,\delta_{ij} - 2\eta\left(D_{ij} - \frac{1}{3}\,D_{kk}\,\delta_{ij}\right) \tag{5.5}$$

where $\eta$ is the coefficient of viscosity defined as the ratio of a certain flux to the force (rate of strain). $K$ is the contribution to the pressure term arising from transport of momentum due to collision while $\bar{F}^*$ is the average force on the blob arising from the intermolecular forces and, hence, is zero. We have shown it explicitly in order to generalize the equations of motion. $J_i$ is the heat flux vector defined by

$$J_i = \sum \frac{1}{2}\,p_i\left(\frac{\mathbf{p}}{m} - \frac{\bar{\mathbf{P}}}{\bar{M}}\right)^2 dN(\mathbf{p}, \mathbf{q}; t) \tag{5.6}$$

and the mean value of $J_i$ is related to the *temperature in equilibrium* by

$$J_i = -\lambda\frac{\partial T}{\partial q_i} \tag{5.7}$$

where $\lambda$ is the coefficient of thermal conductivity. $L$ is again a contribution to the heat flux arising from thermal flow due to collision (see, for example, Ref. 13, Chapter 16).

As has been observed in Ref. 2, equations (5.2) to (5.5) describe the laminar motion of the fluid under consideration wherever the fluctuation theorem is valid in which case we can remove the bar

---

†We here take **q** to be the macroscopic coordinate.

over the quantities $P$, $M$, and $Q$ and put $\mathbf{F}^* = 0$. However, when fluctuations cannot be ignored, we have seen that the equations (5.2) to (5.5) are not characteristic of the type of motion suffered by the fluid. The fluctuation of the force $\mathbf{F}^*$ yields a non-negligible contribution to both the momentum correlation and its mean square value as is evident from equation (5.1). In fact, (5.4) shows how there is a contribution to the mean value of the local energy arising from the average value of the product $F_i^* v_i$. These results clearly prove the existence of a non-zero force arising from an irregular alignment of the intermolecular forces. It is this force that disturbs the local equilibrium and is responsible for the instability of the macroscopic motion. Hence, it is quite reasonable to regard (5.2) to (5.5) as equations describing the fluid in a state of unstable motion provided we remove the bars and view the set of equations as stochastic equations, the stochastic nature arising mainly from $\mathbf{F}^*$. There are also stochastic features in $P_{ij}$ and $J_i$ to which we shall return presently. The terms $K$ and $L$ can be set equal to zero by incorporating its effects in $P_{ij}$ and $J_i$.

Next we notice that the transport coefficients $\eta$ and $\lambda$ defined as the ratio of certain fluxes to certain forces have no meaning in a turbulent flow when all these quantities are defined only statistically (i.e., in a nondeterministic manner). In fact, the introduction of the random force $\mathbf{F}^*$ induces random properties into the transport coefficients. Thus we are led to the inevitable but logical conclusion that the transport coefficients are stochastic variables. In fact, they are stochastic functionals of the forces and fluxes. We can sum up our discussion as follows:

1. The turbulent flow satisfies "generalized" Navier–Stokes equations.

2. These equations are stochastic in nature and in form they appear to be the same as those satisfied by laminar flows of fluids except that the momentum equation contains a random force term of microscopic origin.†

3. The response and transport coefficients are stochastic functionals of the forces and the fluxes.

It should not, however, be imagined that the solution of these

---

†The term $2\{u_i(\mathbf{q})F_i^* - F_i^* v_i\}$ arising from the force term may be incorporated into the thermal flux arising from intermolecular force and, hence, the stochastic nature of the thermal transport coefficients will describe the contribution due to the coupling of the force and the momentum.

stochastic equations is going to be an easy task. What we have done amounts to transfering all the difficulties that were envisaged in the formulation of all macroscopic quantities in terms of weighted stochastic integrals and their consequent connection to the product densities of various orders, into a small number of stochastic equations which appear deceptively simple. Of course, this is one extreme view. An optimistic way of looking at the problem consists in making a suitable guess about the stochastic nature of the force F* and the response coefficients. We believe that the latter view may be helpful in obtaining all the relevant information that have been and can be directly measured by experiments.

Since the generalized Navier–Stokes equations are not very convenient to handle in its most general form, it may be worthwhile to make further approximations to simplify the equations and deal with certain models. Burger's one-dimensional model is a very good candidate for such a study. Flow through a pipe where cylindrical symmetry simplifies the equations can be studied. In fact, some work has been done on this line in connection with the study of incompressible flow through a pipe[17] subject to random pressure fluctuations. These results are being extended to the case of a compressible gas and will be published elsewhere.

## 6. TURBULENT MOTION OF A COMPRESSIBLE GAS: RANDOMLY FORCED SOUND WAVES

We shall finally deal with the turbulent motion of a compressible fluid and consequent generation of resonating sound waves. In this case, the Reynolds number is in the region corresponding to a highly developed state of turbulence and, hence, the perturbation solutions for the product densities obtained in section 4 are not applicable. Hence we have to use the generalized Navier–Stokes equation obtained in Section 5. Thus if $\rho$ is the mass density, $P$ the momentum density, and $Q$ the local energy density, we shall assume that the turbulent compressible fluid satisfies the equations

$$\frac{\partial \rho}{\partial t} + \frac{\partial P_i}{\partial q_i} = 0 \tag{6.1}$$

$$\frac{\partial P_i}{\partial t} + \frac{\partial}{\partial q_i}\left(\frac{P_i P_j}{\rho} + \mathscr{P}_{ij}\right) = F_j^* \tag{6.2}$$

$$\frac{\partial Q}{\partial t} + \frac{\partial}{\partial q_i}\frac{P_i Q}{\rho} = -\frac{\partial J_i}{\partial q_j} - \mathscr{P}_{ij}\frac{\partial}{\partial q_j}\frac{P_i}{\rho} \tag{6.3}$$

where $\mathscr{P}_{ij}$ and $J_i$ can be expressed in terms of the transport coefficients $\eta$ and $\lambda$ (which are random functionals in the present case) and temperature gradient and rate of strain tensor. From (6.1) and (6.2) we obtain

$$\frac{\partial^2 \rho}{\partial t^2} - a_0^2 \nabla^2 \rho = \frac{\partial^2 \varphi_{ij}}{\partial q_i \partial q_j} - \frac{\partial F_i^*}{\partial q_i} \tag{6.4}$$

where

$$\varphi_{ij} = \mathscr{P}_{ij} + \frac{p_i p_j}{\rho} - a_0^2 \rho \delta_{ij} \tag{6.5}$$

$$\mathscr{P}_{ij} - a_0^2 \rho \delta_{ij} = -2\eta \left( D_{ij} - \frac{1}{3} D_{kk} \delta_{ij} \right) \tag{6.6}$$

$D_{ij}$ denoting the rate of strain tensor; $a_0$ is the speed with which sound waves will be propagated in an undisturbed medium when hydrostatic pressure alone is taken into account. Since $\eta$ is a random functional and $F^*$ a random force, (6.4) describes the propagation of sound waves randomly forced. It is also easy to visualize how resonating sound waves are produced by (6.4). Thus, (6.4) becomes a stochastic wave equation and all the relevant statistical properties of wave propagation can be obtained if we are given the statistical properties of $F^*$ and $\eta$. In a certain sense, the problem is essentially linear and, hence, can be solved to a fairly good degree of approximation. In fact some of the methods developed by Keller[18] and the author and his collaborators[19] can be used with great advantage.

Next we observe that the equations (6.1) to (6.3) will assume a simple form if there is cylindrical symmetry. Such a symmetry is realized when we deal with a compressible gas flowing through a circular pipe. Detailed calculations pertaining to pressure and momentum correlations are in progress and will be published as soon as the calculations are completed.

## 7. CONCLUDING REMARKS

In this paper we have examined the kinetic theory of fluids with a view to obtain some insight into the origin of instability that sets in whenever certain macroscopic variables of the fluid are gradually increased. Of course, though we have confined our discussion to the case when the macroscopic density is gradually increased, the arguments used here are applicable quite generally. The kinetic theory that has been formulated in terms of product densities defined in the phase space has one decided advantage, in that it directly yields the moments of the macroscopic variables. It is precisely this situation that has enabled us to examine the fluctuation and predict the onset of instability under certain conditions. In fact, the arguments advanced in Section 3 to explain the manner in which the instability will make itself manifest is based on the approximation law (3.1) in which we replace $g(q_1, q_2)$ by $\chi$. In this connection, it is worthwhile to recall the following observation of Chapmann and Cowling: "The quantity $\chi$ is equal to unity for a rare gas and increases with increasing density, becoming infinite as the gas approaches the state in which the molecules are so closely packed together that motion is impossible." We have connected $\chi$ directly with the fluctuation of macroscopic variables so that the unbounded nature of $\chi$ implies the phase transition that the fluid undergoes be it in a state of rest or motion. Since $\chi$ is only a property of the second-order product density it would be proper to expect the unbounded nature of $\chi$ to be a consequence of the generalised Boltzmann equation satisfied by the product density. If this is proved rigorously, we then have a sound kinetic basis for the phase transition.

As we have demonstrated in Section 4, turbulent motion is a different type of instability in that all the macroscopic variables suffer appreciable fluctuation as contrasted with phase transition where the density alone fluctuates, other quantities like temperature and pressure remaining stable. In this connection it may be worthwhile to explore whether physically interesting macroscopic quantities remain free from fluctuation even when the fluid is in a state of turbulent motion.

In the case of shock waves, the present approach may be useful. Motion of a gas in the shock region is an excellent example of a nonequilibrium configuration and the observed fluctuation of temperature, pressure, and density in any particular situation may be used to determine the correlation functions of the first few orders. In this case the higher-order product densities are functionals (perhaps nonlinear) of the first two orders and it is the author's conjecture that certain macroscopic quantities are stable and free from fluctuation even in the shock region.

Next we wish to observe that the arguments presented here for the transition from laminar to turbulent flow are equally applicable for a plasma. There is a general feeling among the plasma physicists (see, for example, Ref. 20) that the problem of instability in plasma due to turbulence can be dealt with in a satisfactory manner only if the corresponding situation in ordinary fluid flow is tackled. Hence it may be worthwhile to formulate the kinetic theory of plasma in terms of product densities of electrons and ions. The second-order product density which assumes the form

$$f_1(\mathbf{p}_1, \mathbf{q}_1; t) f_1(\mathbf{p}_2, \mathbf{q}_2; t)\left(1 - \left(\frac{e^2}{kT}\right)\frac{e^{-k_D|\mathbf{q}_1-\mathbf{q}_2|}}{|\mathbf{q}_1 - \mathbf{q}_2|}\right)$$

in the case of a stable plasma must be shown to have some properties similar to the case we have dealt in Section 3. More precisely we have to show that it is possible to arrive at *a generalized Fokker–Planck equation for a second-order product density which predicts the instability, which in turn breaks the shielding and disturbs the plasma oscillation thereby allowing the particles to move in a random manner.*

In conclusion, we wish to remark that the contents of the present talk is primarily based on the *idea that all macroscopic phenomena are but consequences of microscopic laws.* Since the equations of gas dynamics are believed to be derivable from some kind of a generalized Boltzmann equation, we have persued this view point logically even to the case involving breakdown of local equilibrium. *Since near equilibrium conditions yield Navier–Stokes equations of motion, the consequences of nonequilibrium configurations cannot be ignored. It is precisely such nonequilibrium configurations that yield different types of instability-like phase transition, turbulence, and shock-waves.*

The author wishes to acknowledge the useful discussions that he had with Professors Alladi Ramakrishnan and S.D. Nigam and Drs. R. Vasudevan, K. Venkatesan, and N.R. Ranganathan.

## REFERENCES

1. M. Dresden: "Kinetic Theory Applied to Hydrodynamics," Colloquium lectures in Pure and Applied Science, No. 1 (1956), Socony Mobil Oil Co. Inc., Field Research Laboratory, Dallas, Texas.
2. S. K. Srinivasan, *Zeit Physik* 193: 394 (1966).
3. P. I. Kuznestov, R. L. Stratonovich, and V. I. Tikhonov, "Non-linear Transformations of Stochastic Processes," Chapter I, Section 6, Pergamon Press, London (1965).
4. S. F. Edwards, *J. Fluid Mech.* 18: 239 (1964).
5. R. H. Kraichnan. *Phys. Fluids* 7: 1163 (1964).
6. A. Ramakrishnan, *Proc. Camb. Phil. Soc.* 46: 595, (1950).

7. M. L. Goldberger and K. M. Watson, *Phys. Rev.* **127**: 2284 (1962).

8. J. G. Kirkwood, *J. Chem. Phys.* **14**: 180 (1946).

9. M. Born and H. S. Green, *Proc. Roy. Soc.* **189**: 103 (1947); **190**: 455 (1947).

10. H. S. Green, "Molecular Theory of Fluids," North Holland Publishing Co., Amsterdam p. 156 (1952).

11. S. A. Rice and A. R. Alnatt, *J. Chem. Phys.* **34**: 2144, 2165 (1961).

12. L. D. Ikkenberry and S. A. Rice, *J. Chem. Phys.* **39**: 1561 (1963).

13. S. Chapman and T. G. Cowling, "The Mathematical Theory of Non-uniform Gases," Cambridge University Press, England (1952).

14. A. Münster, *Rendi, Scuola. Intern. Fis. "Enrico Fermi"* Corso X, p. 81 (1959).

15. C. C. Lin, "Statistical Theories of Turbulence," Princeton University Press, Princeton, New Jersey (1961).

16. S. K. Srinivasan and R. Vasudevan, Nuovo Cimento, **41B**: 101 (1966).

17. S. K. Srinivasan, S. Kumaraswamy, and R. Subramanian, (to be published).

18. J. B. Keller, "Proceedings of the Symposia in Applied Mathematics," American Mathematical Society, Vol. 16, p. 145 (1964).

19. S. K. Srinivasan, Proceedings of Matscience Symposia, Vol. 5 (1966) (to be published).

20. D. C. Montgomery and D. A. Tidman, "Plasma Kinetic Theory," McGraw-Hill Book Co. New York, p. 179 (1964).

## NOTES ADDED IN PROOF

After these lectures were delivered, the author came across a paper by Nordsieck, Lamb, and Uhlenbeck [*Physica*, 7: 344–360 (1940)] dealing with the theory of cosmic-ray showers wherein the authors have made the following observation:

"The theory connects $F(E_0, E, x)$ (mean number of particles) with the known probabilities for the elementary processes through an integral differential equation, which is quite similar to the well-known Boltzmann equation in the kinetic theory of gases. In fact, the problem of finding $F$ is analogous to the problem of how the Maxwell-Boltzmann distribution is reached in time.

When the incident electron has again the energy $E_0$, one must determine the probability $P(E_0, N, x)\cdots$ Theoretically the problem of finding $P(E_0, N, x)$ has some relations to certain little investigated questions in kinetic theory, which may be formulated as follows. Suppose the average behavior of a gas which is not in the equilibrium state, is known; what is the fluctuation around this average behavior? Or we may ask: What is the fluctuation around the hydrodynamical equations of motion? More specifically, this problem implies the question of the calculation of the fluctuation of rates of change or rates of transport in a gas."

The contents of these lectures provide a quantitative connection anticipated by these authors.

# Author Index

# Subject Index